A Portrait of the Scientist
as a Young Woman

A division of The Daily Wire
www.dailywire.com

A
Portrait
of the
Scientist
as a
Young Woman

Lindy Elkins-Tanton

WILLIAM MORROW
An Imprint of HarperCollinsPublishers

HarperCollins books may be purchased for educational, business,
or sales promotional use. For information, please email the
Special Markets Department at SPsales@harpercollins.com.

FIRST EDITION

Designed by Nancy Singer

Library of Congress Cataloging-in-Publication Data has been applied for.

ISBN 978-0-06-308690-6

22 23 24 25 26   LSC   10 9 8 7 6 5 4 3 2 1

CONTENTS

A Portrait of the Scientist
as a Young Woman

CREATING A MISSION TO SPACE

W hat happens if it blows up?" a high school student recently asked me. "Do you get a do-over?"

Ah, no. There are no $800 million space mission do-overs in my world. At the moment of an explosion, we are finished.

In August 2022, if luck, sweat, and the fates align, our rocket will launch our spacecraft off the Earth and begin its 3.4-year journey to the asteroid (16) Psyche. Each asteroid is assigned a number at the time of its discovery, and Psyche was the sixteenth that we human beings have discovered, of the perhaps 1,500,000 asteroids in the asteroid belt. There are not many kinds of solar system objects left to be explored, and Psyche is one of them. We think that it is made mainly of metal. We think—though we don't know for sure—that Psyche may be a piece of one of the first metal cores to form in our solar system. This is why we are going. The molten metal in a very early, hot little planet, called a planetesimal, would have sunk to the interior and formed its core. Then, this planetesimal was broken into pieces rather than incorporated into the Earth or the other big planets of today. Psyche may be part of that exposed core.

Humans have gone to the Moon, and we've sent robotic spacecraft to investigate other rocky planets (Mercury, Venus, Mars), gas and ice giant planets (Jupiter, Saturn, Uranus, Neptune), and icy asteroids and moons (Ceres, Enceladus, Europa, and a couple of comets), but there is one category of solar system object that we have never investigated: the small category of asteroids that we think may be made primarily of metal.

Earth's core and the cores of the other rocky planets are metal. Earth's core is also the source of our magnetic field, the field that may help protect our atmosphere and keep our planet habitable. Metal is clearly a fundamental building block of our habitable planet. But we will never see our own core. To see a metallic world, we think we need to go to Psyche. We won't know with certainty until we get there whether Psyche is a part of a planetesimal core, but Psyche is the only large asteroid in our solar system that is likely to be metallic.

Our solar system started as a vast sparse cloud of dust and gas that was hit with a shockwave, probably the shockwave of a nearby star exploding into a supernova. That supernova was the trigger for our birth. The shockwave compressed the cloud and part of the cloud began to collapse under its own gravity, spinning, forming disks with concentrations of dust and gas in their centers, growing to become young stars. In the spinning disk of plasma, dust, and gas around the nascent star, the first solids formed, little pebbles of minerals rich in calcium, aluminum, and titanium, the elements that condense to solids at the highest temperatures as the hot disk is cooling down from being plasma or molten. These so-called calcium-aluminum inclusions are found inside meteorites that fall to Earth today, tiny survivors of the rapid, violent process of material being crushed into bigger and bigger clumps by pressure and shockwaves in the spinning disk, and the clumps, some as big as continents, colliding and sticking together to form planets.

Those little calcium-aluminum inclusion pebbles formed 4,568 million years ago, and they mark what some call the beginning of our solar system, long before the Earth had formed. If we imagine our 4,568-million-year-old solar system as a 24-hour day, within the first 20 seconds of that 24-hour solar system—about a million years after those first pebbles formed—rock and metal had already collected together into planetesimals, just tens to hundreds of kilometers in diameter. These are the continent-size clumps.

One of these planetesimals, we think, was the parent body of the asteroid Psyche. We believe that Psyche, now an asteroid orbiting between Mars and Jupiter, is the remnant of a battered planetesimal. But just as the length of time between ourselves and the original planetesimals makes seeing that early population of clumps impossible, the distance between Earth and Psyche makes little Psyche, just 138 miles in diameter (the width of Massachusetts without Cape Cod, or the north–south height of Switzerland), too small to see clearly from Earth. The difference is that while we can't go back in time, we *can* bridge that gulf of space. We can send a robot to find out what Psyche really is.

On that day in summer 2022, the efforts of 800 people, 11 years, 2,000 pages of proposals, perfect wires and perfect software and perfect solar panels and bolts and joints and struts, will be perched on the top of about 20,000 kilograms of explosive propellant, and our spacecraft will be sent off into the void. We think it will all work. We are doing everything in our power to make sure it will. We hope it will work.

But, it might fail.

Space is hard.

There aren't many people who have led deep space missions and each of us followed our own path. Indeed, there may be no one single way a person gets to lead a space mission. But if there are a series of paths that have worked in the past, I can say with some certainty that I did not follow any of them.

Chapter 1

ALL I HAD WERE QUESTIONS

The clearest image in my mind is the clump of young poplars at the foot of my parents' driveway, an unpruned, un-thinned clump that concealed a deep opening to the culvert that ran under the end of the drive. Their young spring leaves glittered and clattered in the sun and wind that afternoon as I walked up the road.

It was 1982, and physicist and Nobel Prize winner Professor Hans Bethe had visited my school, Ithaca High, earlier that day and given a lecture about the nuclear arms race and the notion of mutual assured destruction between the U.S. and the U.S.S.R. That dull sidewalk, the colorful cars, and the unruly clump of poplars with shining leaves waving in the wind: Suddenly they were no longer perpetual. They could be annihilated in a firestorm in a second and we would not even have a warning. What could it mean to live in a world so fragile? The fear I already had within me had just discovered a cause, and it leapt up and darkened my joy and my future and took on the name "nuclear war."

Up until that day, I was an earnest, proto-intellectual teenager without a clear idea where I wanted to go next. I was a decent flute player, and I thought about attending a conservatory, but I knew I would never have the talent to be a soloist or first flute in a good orchestra. I was interested in literature and writing and art and art history but I felt those were more my brother's and my father's areas. I was on the fringes of a small group of high school scholars and activists named The Young Plumbers. We wrote a high-quality, regular newsletter of politics and criticism, some aimed at the school board, some at national and international current events. We took advanced placement classes together, notably our English class. In that class we mainly discussed sex and politics. We read *Jude the Obscure* and Shakespeare. One day we were asked to recite a poem of our choice. I was excited: I really did have a favorite poet, Czesław Miłosz. I prepared and memorized my favorite poem, "Study of Loneliness," which ends:

> *If I am all mankind, are they themselves without me?*
> *And he knew there was no use crying out, for none of them*
> *would save him.*

My recitation was met with some silence, and then a cutting remark from the teacher made it clear I was being a show-off. Her comment instantly washed away my pleasure and comfort in the moment with a freezing sense of shame. Obviously, I immediately realized, it would have been way cooler to recite "Jabberwocky." I sat down.

Despite my friendships and activities, moments like this gave me the uneasy feeling that I did not understand how others perceived me, and therefore, that I was not the master of how my actions and intentions were received.

At home, meanwhile, something in my mother had broken. For years she had been the manager of my father's office, perhaps not what her white-glove Philadelphia family had had in mind. Now she came home from the office every evening in a rage, kicking the front door shut and making the window glass rattle through the whole house. She'd fling a bag of groceries on the kitchen counter, stride to her bedroom, and slam the door behind her. I did not see her face or hear a word from her for months. My brothers had long since left for college, so it was me and Dad at home alone in the evenings. We'd watch television and share a quiet companionable pot of tea, and I felt there was a little pool of light in that corner of my life.

Dad was only a shallow harbor, unfortunately. His own anger came unexpectedly and in an instant would burn like the cold of unprotected outer space. A few years earlier, on a family driving vacation to the mouth of the St. Lawrence River in Canada, a single wrong comment had sent him into that uncontrollable void and he had spent the entire trip in a silence so complete and so furious it made my ears ring with the sound of my panicked heartbeat. He drank beer without pause, throwing the empty cans out the window. This from a man who loved the outdoors, and in better days had taught me to canoe, and to identify birds, up in the Montezuma Wildlife Refuge at the north end of New York's Cayuga Lake. On this trip, I sat next to him in fear, all of us jammed together in the car—my mother and eldest brother, Jim, in front, and Tom, Dad, and me in the middle in back. Tom tried light conversational topics from time to time, his voice calibrated to sanity, safety, calm.

Back at home we grew tomatoes together, and Dad showed me how to fix the lawn tractor. We had a tradition for the first ripe tomato of the year. That first bite of the first fragrant, sun-warmed tomato had to be the freshest bite possible—that is, the fruit had to be bitten while still on the vine. This involved worming on my side through the earth of the vegetable garden to get in position, and then shutting my eyes while Dad ceremonially shook on some salt; the salt bounced off the taut red skin, and then I would take my bite.

The love of those moments, and the fact that when he came home from work he called out, "Hello, pal!" brought a tentative feeling of safety to our evening tea and television watching, but I was always careful what I said. I understood that my father was not my protector, but in that period, and later, we could keep friendly company.

When the time for college applications came, my guidance counselor gave me a career questionnaire. I remember that forestry was one of my suggested areas. I was not sure what forestry was. I was interested in science. There I felt solid. I was interested in the animals and plants and landforms of Montezuma and the Finger Lakes, and in animal behavior, and in wildlife conservation, and biology, and also in Earth science.

The more I thought about geology, the more I felt calm and comforted. I knew the solar system was over four billion years old. What was the matter of a second, when the planets orbited and the Sun shone for billions of years? What was the preoccupation with a fear, even the fear of nuclear destruction, when that fear was locked within one small human, living for just a cosmic moment, on one small planet that itself was going to go on for billions of years more? That geologic timeline spooling out and out into the past and then again into the future felt like a tall cool drink on a hot day.

"WHAT'S THE INTEGRAL OF 1 OVER X PLUS 1?"

I had come to Frank Morgan, the ultra-charismatic young professor teaching my freshman calculus class, because I was struggling. I had walked to his office with its severe black door in the echoing limestone halls of the MIT math department with a sense of dread, but without other ideas for how I was going to solve my problem. Now, instead of answers, he had instantly fired back another question in response.

I was getting a C in calculus and to me that felt like an F. I had never had to study in high school. Despite that, when my high school precalculus teacher, whom we all worshipped, agreed to write me a recommendation letter for MIT, he assured me, "You'll never get in." Needless to say, the standard MIT freshman load of physics, calculus, chemistry, and a humanities class (for me, art history), plus intense socializing in the dorms, strained my weak study skills to the breaking point.

Professor Morgan waited silently for my response. I felt my vision contract as I searched inside myself for the answer, which I did not find. My hands felt cold. I mumbled something about natural log, but it wasn't a confident answer, and he wasn't about to tease it out of me.

"That's why you're not doing well," he said. "You don't study hard enough."

I HAD ARRIVED AT MIT FEELING STRONG AND CONFIDENT AND VERY much myself. That summer, I had bicycled, alongside my friends Theo and Chris, from Madison, Wisconsin, over the Great Lakes through Canada, and down through Ontario back to Ithaca. I had even built my own bike from individual components. But this experience in calculus was an early sign that MIT was going to be a different beast from high school. I quickly perceived that I would be hard-pressed to fit in as an explorer or field geologist, though that's what I wanted most. I had arrived that August so full of confidence and hope; they were soon knocked out of me. I now had only questions, no answers. But at that time, questions were a way of stretching out my arms in the darkness, of trying to understand the landscape.

How should I compare myself with the student who confidently told us all he spoke the purest, best version of English (from Cambridge) and the purest version of French (Orléans), and the best version of Hindi? What about the students who only really spoke to the other students who studied computer science? What did I do with all these hyper-confident people who made me feel like I was in a different class of humanity? I had a provisional answer: avoid them.

I had grown up reading the stories of the great explorers. I read *Endurance* and in my mind I traveled along with Shackleton and his team as they survived the icy imprisonment and eventual destruction of their ship, and the necessary—and hopefully temporary— abandonment of the less able people on Elephant Island with just a freezing ledge of rock for shelter. Shackleton and just a few companions made the open sea voyage to South Georgia Island and completed the near-miraculous rescue of all his men, all alive. Carveth Wells took me to the jungle with him in *Six Years in the Malay Jungle*. He introduced me to durian fruit, and orangutans, and long-term, dedicated travel. Perhaps my favorite, and most deeply experienced, travel adventure was Ivan Sanderson's *Animal Treasure*. I went with Sanderson to West Africa in search of rumored animals, not yet confirmed to the European and American scientific establishment, like the African clawed frog, and the potto, and the Giant Booming Squirrel. I learned about malaria and about searching through the night after mysterious yelps and quacks, trying to find their animal source. I imagined entering such a state of fearlessness that I would thrust my arm into a hollow tree in the middle of Africa in the pitch-black night.

In these books I witnessed courage and discovered the thrill of the unknown. What I did not yet understand was that I was not invited. Not just because I was young, but because I was a woman. Growing up in egalitarian, even utopian, hippie Ithaca, I thought I was allowed to do all these things. I thought the world of exploration was for me, too.

But as my conversation with Frank Morgan had begun to show me, I would have to earn my place here; in 1983, my admission to MIT was not nearly enough for me to be truly accepted into the world of

science and exploration. At the time, the MIT undergraduate population was only about 20 percent female, though the fraction of women
was on the rise. MIT was in transition. Many of us were in classrooms
with no other women, or just a few. Almost all of our professors were
men. Almost all the revered figures shown in bas-relief on the campus
buildings, or in oil paintings or murals, were men.

Why is being a woman in science important? Not, perhaps surprisingly, specifically because I am a woman. But because my identity
and my appearance bring to me certain biases and certain privileges;
and so does my life story, just as each of our stories do. It's also true
that for centuries, science was a man's world, and in the West, a white
man's world. Most universities over the last thousand years excluded
women. The work of female technicians, who were in those positions
because more senior positions were not open to them, was credited to
their senior male colleagues. Academic leadership remained almost
exclusively male until quite recently. That has begun to change in
recent decades, but the sense of the scientific community as a male
domain is still strong. It certainly felt that way at MIT in the 1980s.

Having made my way, haltingly, through the physics, chemistry,
and calculus freshman requirements, it was time to declare my major.
I thought yearningly of animal behavior, which I loved from my years
riding hunters and playing with every kind of pet, and watching wild
animals and birds (and dreaming my way through all those books of
the great explorers), but I was, embarrassingly, stymied by organic
chemistry. Where were the universal comprehensible rules of molec-
ular interaction and reaction?

One mediocre grade in a freshman class, and that path, in my
mind, was now closed. I turned to the topic that felt more logical: I
declared geology as my major.

The undergraduate population in my department, Earth, Atmospheric, and Planetary Sciences, was mostly female. A geology major

was viewed by some as a refuge for people who couldn't make it in the "hard" sciences, or engineering. This is a false assumption; excellent Earth and planetary science scholarship requires math, chemistry, physics, and biology. At that time, though, and in some universities to this day, geology is viewed as easier than other science majors, although the content is no easier than that of any of the others. Earth science, however, had lagged behind biology, which in turn had lagged behind chemistry and especially physics, in maturing as a field.

In the early 1800s, Michael Faraday made fundamental discoveries about electricity and magnetism in a simple laboratory; he could have done them in his kitchen. Around the same time, Charles Lyell published his *Principles of Geology*, which contained the first clear, supported argument that geological processes are slow and uniform and extend over far greater reaches of time than had been previously imagined by Western natural historians and philosophers. Some had thought the world had been in existence for only a few thousand years, while others posited ages in the tens of millions, none longer than about 400 million years (Lord Kelvin), still only one-tenth of the Earth's real age. Lyell drew upon travels and observations from all over the world to make his arguments, which spanned three volumes. Without that understanding of the immensity of time, geology made no sense, and could not be studied as a science. Lyell's work greatly influenced Darwin, and helped geology begin its long, slow climb from descriptive science to hypothesis-driven science.

Already, as a freshman, I had the idea that the real goal was doing, not studying about. MIT is dominated by its research culture, and has as many graduate students, who spend all their time doing research, as it has undergraduates. It's normal, almost expected, for undergraduates to have research jobs with faculty. Sitting in the kitchen area of our dorm hallway in the late afternoon after classes, I watched some students trail in after sports practice, and others trail in after time in their labs. Finally, at the end of the first semester of my freshman year I summoned courage that surprises me to this day, and called the famous professor Nafi Toksöz. Nafi came to MIT in 1965, the year I was born, and founded the Earth Resources Laboratory, where he and his team made fundamental discoveries about earthquakes, plate tectonics, and natural resources. He could have stood on his high pedestal and been intimidating, but instead, he greeted the world with a warm, humble smile, I soon learned. On the phone I asked him if he had a research position for a freshman with no experience of any kind. Some experience, actually: I had been a post office clerk.

Nafi took my call, and listened to me, and he hired me and set me to work writing code for the New England seismic network. These seismometers, scattered around New England, were used for monitoring earthquakes, and also for detecting Soviet nuclear tests. We could tell the difference between the two by the shapes of the waves recorded by the seismometers. So now I was both an Earth scientist and doing something about that terrifying nuclear world of Hans Bethe.

For hours each day I would sit in the Earth Resources Lab building, surrounded by computer terminals and the great big reel-to-reel tape machines of that era. The next room had an under-floor cooling system for the big computers, and we sometimes put sodas under the floor tiles to make them cold. Our room was filled with the sound of key taps and tape spooling, the tables covered with continuous-sheet printer paper with those perforated edges. There I worked with a couple of undergrads and a couple of grad students and a couple of staff scientists, led by the kind, smiling, and supportive Mike Guenette. Mike taught me how to code in FORTRAN and set me to work writing a program that located the town closest to the coordinates of any given earthquake our New England network detected. That team of kind, supportive people became my home port at MIT. We went to dinner at Mary Chung's in Central Square and ate the searingly hot Suan La Chow Show. We shared our lives in that magic way which happens most often at that age.

Since I lived on campus not far from the lab, I often drew evening duty changing the paper rolls on the seismographs. I felt strong and even a tiny bit important as I walked, holding the building key in my pocket, across dark Ames Street and over to the Earth Resources Laboratory. I loved turning on the lights in the dark room and knowing how to stop the recording, and remove and replace the foot-wide cylinder of paper with its many seismic wiggles.

One night I turned on the lights to discover that the tiny pen had drawn huge waves across the heat-sensitive paper, waves so large that they crossed over all the other wiggly lines and bumped against the limit of the pen's available movement. There was a literal red phone for such emergencies, which I picked up that evening for the first time to call the lab manager at home and tell him that a big earthquake had just been recorded. No one yet knew where. This was the giant 1985 Mexico City earthquake, which showed up as huge rolling waves and spikes on the seismic monitoring network all the way into New Hampshire.

Here I had found a new team that felt like family. This group and the work we did together gave meaning to what I learned in the classroom, and unlike completing problem sets, achieving goals with Nafi's team felt real. I began to feel I had found my footing a bit, as I completed the year and went home to Ithaca for the summer.

In my sophomore year, beginning the studies of my major, I was in a class called Igneous Petrology. Our professor, Tim Grove, was serious and rigorous and organized with pages of notes, but also approachable: he wore Birkenstocks to class with his proper shirt and trousers. He was rigorous, though, and would look at us unsmilingly through his wire-rimmed glasses when we fumbled an answer.

One day Tim announced he was looking for an undergraduate to work on some research. I eagerly volunteered. Later that week I met Tim in his office on the twelfth floor of the tall MIT Building 54, more commonly known as the Green Building. The office had a microscope, a computer, piles of papers and books, and Tim's modest metal desk. The window had a sweeping view over the Charles River to Boston, but Tim's blind was often down. Tim told me he had an idea that was risky; the required experiments would be time-consuming and might or might not work, so the project was perfect for an undergraduate, who would learn all about scientific process and how to do experiments, without any downside if no results materialized. High risk, high reward? No surprise and no problem if the project failed in the end? I was *in*. I started asking questions. Tim answered a few, and then said, coldly, "Questions are fine as long as there are not too many of them." I felt a chill, and realized it came from a memory of my father, and his temper, as sudden and irreversible as a fall into a chasm, requiring just one wrong step.

Tim introduced me to what became my first great scientific obsession: conducting high-temperature and high-pressure experiments that mimicked the interiors of the Earth and the Moon, and allowed me to interpret what the temperature, pressure, and composition of those interiors are now, and what they were in the past. That felt like real science to me, teasing out truths of our universe from seemingly unrelated bits of rock, and learning about places too remote for us to ever visit. I could work at the processes and become an expert. I could learn something about the Earth no one else knew.

Tim's project concerned the most common rock-forming mineral on Earth, feldspar. In many igneous rocks—rocks that have solidified from a magma—two feldspars coexist: one with high potassium, and one richer in a mixture of calcium and sodium. The exact compositions of each—that is, their relative proportions of potassium, calcium, and sodium—vary with the temperature and pressure at which they formed. Tim thought we might be able to heat powdered rock at set pressures and temperatures, extract the resulting rocks, and measure the potassium, calcium, and sodium in their feldspars. By doing this over and over at different pressures and temperatures, we would build a database of the compositions of feldspars formed at a range of pressures and temperatures. Then, anyone with a natural rock containing two feldspars could measure their compositions, and compare them to our database to discover the pressure and temperature of their natural rock's formation. This kind of calibration is called a geobarometer and a geothermometer. We were setting out to make the first feldspar geobarometer and geothermometer. Scientists could take rock samples and find out the temperature and pressure of their formation, answering questions like, how deeply was this rock buried in the crust before it was thrust up into a mountain?

Tim explained that the furnaces we needed for our experiments didn't exist in his lab yet, and we would build them together. For several months, I was Tim's apprentice. In one room of the lab, tables and old soapstone counters were filled with dissecting scopes, squat glass evacuated storage bells, polishing sandpapers, acid bottles, boxes of gold tubing, a cardboard barrel of barium carbonate (not for rat poison, but for a special kind of high-pressure experimental apparatus), welding machines, and hot-pot furnaces over which teetered shelves of heavy books. In another room, the high-pressure and -temperature furnaces were packed as close as could be designed. To get to the back of the lab where Tim and I would work, I had to turn sideways and slide between the lab bench and the big disk of the pressure gauge for one of the furnaces.

Tim followed that most correct and fundamental rule in science: You must know how your tools work. You should build them yourself. When he hired me, he looked at my hands: Was I a tool user? Yes, I was—the calluses from building and riding my bike were still there. He showed me how to connect high-pressure tubing from the water reservoir to the valve in front of each bomb and furnace, and how to connect the valve to the bomb's screw cap to that thinner, flexible high-pressure tube, itself as thin as a wire.

We built a dozen cylindrical high-temperature ovens, lying on their sides and opening like clamshells, into which would be laid smaller metal cylinders—the cold-seal bombs—each one about the size of a thin baguette. Each bomb had a tube drilled into its center, where we would insert a tiny volume of powdered rock welded into a gold capsule about a centimeter long. One end of the bomb was closed, and the other end had a finely machined screw cap connected to one of those wire-thin metal water tubes. The metal tubes connected in turn to a high-pressure water system. The metal bombs were pressurized from within with that water, and they lay in furnaces as hot as 1,650 degrees Fahrenheit.

In the vernacular of today, what could go wrong?

Those high-pressure fittings were fussy. They had to be tightened just the right amount and not too much. Tim showed me how to read the pressure gauge for the water reservoir; if it was falling, there was a leak somewhere. Sometimes the leaks were pinholes in the high-pressure wires, and extremely hot, pressurized water would spurt out in an almost invisibly fine stream. Everyone in the lab—myself and the other students—was aware at all times of the danger of the furnaces. Certain kinds could explode, and pressure lines could open, material could burn, and the metal parts of the furnaces could crack under the pressure with concussive bangs like gunfire. There had been all sorts of disasters, both small and large, though because of our training and care and some luck, no one had ever been hurt. Tim told me to call out "Noise!" in the moment I dropped a tool, so that its clatter on the floor would not cause too bad a startle for the others in the lab, all tuned to a high pitch by the challenges of the apparatus.

When we were ready to do the first round of experiments, Tim explained how to request pure mineral samples from the Harvard collection to supplement what he already had in the lab. These mineral names still read like poetry. There was Amelia albite from Amelia Court House, Virginia; Crystal Bay bytownite from Crystal Bay, Minnesota; Lake Harbour oligoclase from Baffin Island, located in the Canadian territory of Nanavut; the Sannidal andesine from Sannidal, Norway; Hugo microcline from the Hugo pegmatite, South Dakota.

I ground them to powder and measured them to the ten-thousandth of a gram in order to make exactly the starting compositions we wanted. Then, I learned how to make the tiny gold capsules, hammering the capsule into the correct width, and welding in the tiny measure of rock powder and droplet of water using an arc welder with a hand-sharpened carbon tip. The welder had an electrical short in it, and sometimes when I stepped on the foot pedal to start the current, the machine shocked me in my eye socket through the welding glasses my face was pressed against. Having a steady hand under these circumstances took special determination.

Finally our first ten experiments had sat baking in their ovens, up at pressure, monitored every day, for six full months. One by one, using welder's gloves, we lifted the bombs from their fiery oven beds. We held them lid-down and rapped them with a great wrench to loosen the experiment within its narrow drilled cavity, allowing the experiment to fall into the cooled head of the bomb, which quenched the experiment to room temperature quickly and without allowing any additional reactions to occur. Each little gold capsule in turn tumbled out of its bomb after the lid was unscrewed. One or two were discolored and wrinkled, and Tim explained that those had almost certainly burst during their long bake and would be useless. The others we carried reverently into the microscope lab. One at a time we would open them under a binocular microscope, each glittering capsule snipped open over a clean petri dish. So much labor had gone into them that we treated each one as delicately as a butterfly, more precious than the gold that wrapped them.

I watched intensely over Tim's shoulder as he clipped the end of the first gold tube and upended it in the petri dish; he was lit like a star actor on the stage by a focused light and enlarged under the dissecting microscope. A drop of water, and a tiny pile of sand poured out. "Ugh," Tim grunted.

"What? What happened?" I asked.

"Nothing," he said. "Nothing happened. That's the same rock powder and water you welded into that capsule six months ago." Well, we thought, perhaps that furnace had a faulty temperature sensor and it hadn't been at a high enough temperature. Our frustration grew as one after the other produced the same disappointing heap of powder, unchanged from when it was first welded into the capsule. Those six months had not been enough; the temperature and pressure had not been enough. No new feldspar minerals had crystallized.

Losing all those experiments was no small failure. The time, the cost, the care . . . in all, the better part of a year was gone with nothing to show. But at the time, I knew no better—this is what experimental science is, I thought! We'll try again, said Tim, at higher pressures and temperatures. And so, we started again, from the very beginning.

By midway through my junior year, thanks to the higher reaction rates at higher temperatures, the experiments were working. When I snipped open a gold capsule under the dissecting microscope a gratifying tiny pebble of feldspar, newly crystallized in the oven from its powdery starting materials, rolled out into the petri dish. These

I mounted in epoxy, polished by hand using increasingly fine sand-papers and wet polishing compounds, coated with vapor-deposited carbon, and analyzed in an electron microprobe.

Oh, that electron microprobe. The instrument filled a whole room, and the operator (me!) sat in front of an angled bank of screens, knobs, dials, and switches. On one side was the big sample cham-ber, held at high vacuum, and chilled with a Dewar of liquid nitrogen that I would refill for each measurement run, often spilling a shower of skittering liquid nitrogen droplets across the floor. A microprobe measures the atomic composition of solid materials by shooting a col-umn of electrons at the surface of the sample. The electron gun is just a hot tungsten filament in the vacuum chamber. The electrons stream off the filament, are focused by electromagnetic lenses, and strike the grounded sample in a beam as small as a micron across, a hundred times smaller than the smallest thing you can see with your eye, smaller even than bacteria. The electrons add energy to the atoms of the sample, causing the electrons in those atoms to rise to higher energy orbits around the atomic nucleus. When each electron loses that added energy, it sinks back to a lower orbit, and releases an X-ray. Those X-rays fly away from the sample, and some of them strike spe-cial crystal sensors in that sample chamber. Each atom releases X-rays of characteristic energy, and so by counting those X-rays, we can know how many atoms of each kind are in the sample.

The audacity to invent such a machine! I wondered at it then, and still do now. The electron microprobe allowed me to measure the ex-act composition of the feldspar minerals that formed in my experi-ments, and, in the end, allowed Tim and me to produce our feldspar geothermometer and geobarometer.

But first, I had to learn how the machine worked. The friendly, brilliant, über-nerdy lab manager who knew every in and out of that machine sat with me for hours and days while I learned and practiced.

First, I watched and listened. Then, I operated the instrument while he watched. Then, I operated the instrument with him outside the room. Finally, after a couple of months, I could fill the microprobe's liquid nitrogen tank; prepare my sample; coat it with carbon (its own exciting and error-prone activity, involving another vacuum chamber, a graphite rod, and electricity); put it into the microprobe; bring the probe sample chamber back to vacuum, calibrate, and standardize the probe so its measurements were reliable and replicable; and finally, measure my own experiments by myself, sometimes in marathon overnight sessions when the instrument cost was lower because of decreased demand (to keep the facility running, user time is paid for on research grants).

Those nights were the first times I felt like a real scientist. The probe room was windowless and refrigerated, and completely dark while operating, so that I could better see the relatively dim images formed by bouncing electrons off the sample. Not only was I measuring composition with a focused electron beam, but I could also "see" my sample using electrons, instead of photons. The electron beam rastered back and forth across the sample to build up an image, and rather than seeing colors in visible light, I was seeing composition using electrons. The denser the atoms, the more electrons were bounced back, and the brighter the image was. Bright = dense, and dark = less dense. I understood the machine, could use it with some confidence, and was building a database of information about feldspar behavior that no one else had ever done. Like a lightbulb with a loose connection, from time to time my sense of confidence would flicker on and shine, and then, flicker off.

ALL OF US WOMEN WERE TOLD AT ONE POINT OR ANOTHER THAT WE were at MIT on sufferance, that we were not really good enough. I was told that first by my high school math teacher when I was applying, and I was told it both explicitly and by implication many times along the way. But feeling desired for my physical body, feeling in demand for a date, that was easy and everyday. When we were freshmen, a male friend confided in me that they had joked behind my back about the "Lindy Lottery," that is, the competition among the undergrad men to go out with me. Please understand, this had nothing to do with my being attractive, brilliant, or vivacious; it had to do with how few women there were and how many men. What was my value? Was it as a girlfriend, or as a scientist? Around this time, a female graduate student visited the home of one of our faculty, whose small child told her, "But you can't be a scientist! You're a girl!"

There was much conversation on campus about the gender ratio at the school. Many men and many women believed, and said out loud, that women were there because of a kind of affirmative action, and were not as accomplished or ready as the men. The admissions office even went so far as to publish SAT scores and other metrics to show that women deserved to be there. Still, the emotional response often felt more compelling than the data.

I needed to know if more of us felt as I did. I approached Professor Lotte Bailyn of the MIT Sloan School of Management about advising me while I created a survey of my fellow undergraduates about their attitudes around women in science, in general, and women at MIT, in particular. Professor Bailyn agreed to advise me, and with the help of Shirley McBay, dean of student affairs, and Lynn Roberson, from MIT's Programs and Support for Women Students, I wrote and distributed a survey, and during my junior year, I learned how to process data in the statistical data program SPSS, analyzed the results, and I wrote a report.

The survey results showed that 46 percent of the women and 53 percent of the men felt that women were preferentially admitted.

Though the Advisory Committee on Women Students' Interests had shown in repeated studies that men and women performed equally well in their academics, only 55 percent of the women and 32 percent of the men were aware that this was so. Thus, a feeling pervaded campus that women did not belong at MIT. I felt the resulting "imposter syndrome" as acutely as anyone; even the all-male statues and bas-reliefs around campus seemed to be frowning. Another tedious stereotype my survey confirmed: 79 percent of women and 71 percent of men agreed that there was a strongly negative stereotype about MIT women, who were most often described with the adjectives "ugly" and "boring."

The few ultra-confident aside, I quickly found that both men and women at MIT often felt like imposters. In the mid-eighties, nerds were not yet cool. In my coed dorm (at MIT dorms act as living groups and support systems for the duration of your undergraduate experience), we had an intense network of friendships and a wild round of activities. We did our homework together, writing on the chalkboards in empty classrooms at night; we brought crates of fruit from the Haymarket home on the T and had daiquiri parties; we muscled through impossible problem sets by working together in groups, often powered by raw chocolate-chip cookie dough.

I lived on Third East in the parallel dorms known as East Campus. My freshman year, the first building for MIT's media lab was going up across the street. Pile-driving the foundations deep into the mud of the Charles River Basin caused coffee mugs to hop across their shelf in my room and smash on the old linoleum floor. The din of construction made sleeping late impossible and drove us away from the dorm for daytime studying. And so, of course, we had our revenge.

The media lab building's famous architect, I. M. Pei, had added three squares of color—black, red, and yellow—to its otherwise white-tiled street face. Our floor had a trademark color, mint green. We started sneaking across the street at night and painting a fourth, matching square, in mint green. During the day, the construction team would take it off. We'd put it back. And then, the night before the grand opening and unveiling of the building, in 1985, we executed an MIT-level hack by painting it back on, under the noses of the guards placed specially for that night. Pei was startled but responded graciously when the mint square was unveiled with the rest the next day.

Meanwhile, the feldspar data was adding up to a beautiful story. Tim invited me to make it a master's thesis and I discovered I had sufficient credits to complete my bachelor's in geology and concentration in women's studies in seven semesters. I finished my graduate coursework and thesis in an eighth semester, and thus received the bachelor's and master's, respectively, in geology and geochemistry together at spring graduation 1987. At the end of the year, Tim submitted our work to a scientific conference, and invited me to be there and present it. My instant reaction was no. I was so terrified by the idea I did not even attend.

Asking questions, in fact, had become fraught. If I asked too many questions, I'd be viewed as weak. Asking questions implied ignorance, unacceptable at MIT. You were expected to have learned it already, or to figure it out on the fly. Asking a question risked revealing to others that you had missed something obvious, and that was a significant risk to run. The one exception to this rule occurred during the weekly seminars given by visiting scientists. The faculty, grad students, and some undergrads would gather in the big room on the ninth floor of the Green Building to listen to some eminent researcher and then prove their intellectual worth afterward, on the battleground of questions and stinging comments.

One elder statesman was famous for waking from his nap, shouting, "That's just poetry!" (a kind of über-insult implying flowery language with no substance), and then going on to eviscerate the speaker with a particularly insightful pseudo-question, that is, a statement phrased as a question but formulated to point out a fatal weakness. Questions were swords, not magnifying glasses.

What I needed was a magnifying-glass question. If I wanted to be a scientist but was not yet ready, what was I ready for?

Chapter 2

IN FRAGMENTS

With my stomach alternately buzzing with nerves and sinking with despair, I drove into a dusty parking lot off a long country road in Great Mills, Maryland. Inside a low brown building I found a small waiting room with a few soft chairs, a radio playing, and the ubiquitous piles of magazines, but no receptionist. The door out of the waiting room and into the rest of the office suite was shut. I checked my watch, and saw that I was a little early. Should I knock on one of the closed doors? That seemed wrong. I decided to wait. I sat down and picked up a magazine, trying to breathe calmly. I wanted to appear normal, while at the same time trying to imagine how I could describe my anxiety and despair vividly enough to get the help I needed. I was placing all my hopes on this therapy, on this person I had never met.

A few minutes later the door to the offices opened, and a woman about my own age stuck her head through. "Lindy?"

"Yes," I said as I rose and walked through the door into Mary's office. Her long hair pulled neatly back, and her round figure dressed casually, she gestured for me to sit on a couch. The room was filled with books and pictures and had a pile of toys in the corner for kids. She sat in her chair at one side of the room and I sat on the couch facing her. I had no idea what was supposed to come next.

She said hello, and she asked me what made me want to come to therapy. I took a breath, looked down at the floor, and then began to tell her my story. I talked about my recent history: the death of my marriage, the beginning of a new relationship, and being plagued by depression. I explained that I didn't know why the depression had returned, since my life seemed to be fixed now, and there hadn't been anything so terrible in my life, nothing that fitted the magnitude of the depression I felt. There wasn't any reason for it.

Mary's first response was, "You seem to have come here to remember, maybe to remember why you are angry?"

While at first this seemed excessively abstract, it also struck something in me and seemed real, or at least, startling and important.

Then she said, "I think you have a fear of the unknown. We have to think about what is unknown."

And that little statement gave me a stab of panic. It stirred something I immediately pushed away.

WHEN I WAS EIGHTEEN AND HOME IN ITHACA FOR THE SUMMER AFter my freshman year of college, I went with my friend Sara to a party. She'd been invited to it by a friend of her roommate's, and I'd been invited by her. We followed the directions to drive along a dark country road, park on the side, and then walk down a dirt road through a dark forest to a stream, Fall Creek. Telling it this way brings up visions of true crime stories, but in Ithaca in 1984 this was just a normal adventure on a gorgeous summer night.

By the stream were a couple of dozen Cornell college students hanging out talking. Some people made up a sweat lodge, a semi-traditional version of a sauna, with their addition of some marijuana, but I didn't do any kind of drugs and declined. So did tall graduate student Curtis, and we sat together and talked. He was getting his Ph.D. in wetlands ecology. We talked about sports. He had been in training for consideration for the Olympic cross-country skiing team, and he was a fencer, and practicing aikido. He also knew something about horses; hunt-seat equitation and jumping was the sport I was most expert in. And he was strikingly handsome, strong, six feet tall, pale blond. He could have been Norwegian. We talked about science. I fell in love that night.

During the fall Curt came to visit me at MIT and we drove up into the White Mountains to go backpacking. Curt tried to identify all the plants around us, a part of his discipline as an ecologist. He excitedly gave me a little leaf to chew, but alas, it was partridgeberry and not wintergreen. The night was cold and our water bottles froze. We loved every minute of the trip. I loved being able to show him I was competent in the outdoors, and tough in bad weather. Everything was easy, everything was fun.

Curtis came from a highly accomplished and prominent family with deep connections: secretaries of commerce, ambassadors, assistant secretaries of state, centuries of investors in American industry, major philanthropists. I came with only a few strong floor beams in my personal house: my academics at MIT, my accomplishments in horse riding and flute playing, and some fairly prominent relatives on my mother's side of the family. For a couple of weeks the following summer Curtis and I visited his family up and down the coast of Maine, taking ferries out to some islands, being picked up in sailboats. On a sail one long glittering afternoon, I was questioned intensely by several men of Curt's father's age as we cruised around Casco Bay. I was a horseback rider. So was one of their granddaughters; very good. I was studying science at MIT. The other man's grandson was at Harvard. His family always went to St. Paul's School. Where had I gone to high school? I went to public school in Ithaca. Ohhhh, he intoned, with raised eyebrows.

Curt had no problem with these seeming failings of my lineage and upbringing. But my own greatest feeling was yearning. How I wanted a presentable pair of parents, and a clean beautiful family house to return to, and predictable, pleasant holidays, and good behavior. I wanted it with every part of myself. That wanting, though, felt far away and irrelevant when Curt and I were together. We hiked, backpacked, canoed. We hosted dinner parties, built things, were happy. The following year, we got engaged.

That summer, we were visiting his parents at their place on the Chesapeake Bay. We were sitting around the table having another of the delicious, healthy meals from scratch his mother produced three times a day without fail; the Brittany spaniel was sleeping under the table, the halyards on the sailboat masts clinked in the wind, and the horses snorted from time to time as they grazed in the pasture out back. We had some savory soup with sherry, and salad from her garden, and make-your-own sandwiches with roast beef, turkey, cheese, various breads and sauces. When we told them our news, Curt's mother, Janet, shrieked with pleasure, and then took the family diamond ring right off her finger and handed it to Curtis, who put it on my hand. There I was, a kid studying science at MIT, wearing jeans and anoraks, with a giant shining heirloom diamond on a big platinum band on my finger. I might be all right after all.

FOR REASONS STILL UNCLEAR TO ME, WHEN I WAS TWENTY-ONE AND I finished my master's at MIT, I was not confident enough to go ahead into a Ph.D. program. Several of my uncles were in business, and I was so interested in what they were doing. What was "business," anyway? I interviewed through the MIT career office and got a job with Touche Ross (now Deloitte) management consulting in Philadelphia.

No one was happy about this except me. My parents thought I had sold out. Tim Grove might have been frustrated with me, or thought my going into business was a waste of time. I'm not really sure what he thought, actually, because we didn't talk a lot in the months after I told him. I, however, was eager to learn what this business world was about, and also to make a little money. Still, I was a bit jangly because of the people around me who had wanted something different for me. Were they right? Or was I?

After a training period with Braxton Associates in Boston, where I learned to use Excel, mastered the market-share bubble chart, and got used to wearing suits, I moved into a little apartment in Narberth, Pennsylvania, and started taking the train in to Center City, Philadelphia, each day, to work in a high-rise. I felt so professional, so in the real world. In the train I'd see other people—people who seemed somehow adults—reading the pink *Financial Times* and I felt I was part of something. In the Center City train station I'd buy crisp toasted cornbread slathered with melted butter, which I'd eat as I walked to our high-rise offices. The other research associate in the office, a beautiful, friendly, stylish young woman who drove a BMW and had a boyfriend at Wharton, and I were assigned to project teams and helped on consulting projects in all the ways we could. Our work included making charts and presentations, getting reports bound, taking data, and even helping to strategize on projects. We were included, and invited to stretch.

One of my first big projects was with Boeing Helicopters. Rather than taking the train in to Center City I drove my beloved tan 1979 Volvo sedan with a sunroof, or moon port, as my brothers and I called it, back and forth to the big Boeing plant south of Philadelphia, down I-95 past the giant oil refinery with its perpetual flames, especially eerie in the dark. This plant was remanufacturing the Chinook CH-47A, B, and C helicopter models into D models. The factory line for giant helicopters was an impressive sight: a huge building filled with stations through which the helicopters moved, surrounded by giant machines; carts of tools; periodic tall fences protecting parts control areas with shelves up to the remote heights of the ceiling; desks; golf carts; and people, people, people. The echoing space was filled with the clang of metal and the smell of machine oil.

Helicopters would be disassembled, and parts would either be scrapped, remanufactured, or cleaned and saved. Every part had to be tracked, and the computer inventory had to match the number in the parts control areas. We were helping Boeing improve their inventories. My job was to track the paperwork. I started with, say, a particular bracket that stabilizes the transmission, and then I interviewed everyone who touched that bracket. *What form do you fill out? What do you put in the computer? What are the options for your next decision? Where does the bracket go, where does the paperwork go, and who do you talk with next?* I built a giant spreadsheet of all this information, and then I drew an even more gigantic flowchart of all the possible paths for parts and paperwork. The flowchart looked like a platter of spaghetti and made obvious the room for improvement. My team loved it, and the client loved it, too. In the meeting where my manager presented these results to the client, the client asked, "Could I keep this printout?" That was *my* printout, my flowchart. I had created information that hadn't existed before and that the client valued. I felt a burst of pride.

I learned that if someone on the line worked fast enough to make his coworkers look bad, he'd get his tires slashed. I learned that it was important to go out and eat wings and drink with your team after a long day. I learned I got drunk easily. I learned to drink less. And then I learned the lesson that changed the way I look at the world.

The managers of our project had taken all the information we had gathered, and they had come up with a solution for Boeing. They had an idea about how the people and the paperwork and the computers could be organized differently, so that the parts would be tracked better. We all worked on how to explain this idea, how to sell it, and then we presented to the Boeing management. They liked the idea. That was it.

Remember that I had just come from four years of intensive undergraduate and master's-level scientific training at MIT, and three of those years I had spent doing some hard experiments: gathering data, analyzing, and interpreting the results. I had come from a world where you painstakingly produce data that measures something, and then you work for weeks or months trying to understand exactly what that data is telling you, to extract the elusive but absolute and immutable truth of the natural world from it. Here, at Boeing, our team had analyzed our data in the same painstaking way, but then we had *made up the answer in our heads*. We invented a process that would fix the problems.

When it comes to altering a team's behavior, or organizing a group of people in any way, there's no absolute universal truth, no physical law. Whatever you make up, if you can convince the people around you that it's right, becomes real. That is not how science works. But it *is* how teams of people work. It felt like magic then, and it still feels like magic to me now. This lesson would stay with me through the years ahead of building and working with research teams.

AFTER A YEAR IN PHILADELPHIA MY KNEE PROBLEMS BEGAN AGAIN. I'd had bad knees on and off since the end of high school, when I had surgery first on one and then the other. At some points the pain had been so bad that any walking was difficult, and before my first surgery, my knee spontaneously dislocated a couple of times. My right knee was rapidly becoming unworkable again. My new surgeon said my knee was filled with scar tissue, and the whole joint lining needed to be removed. Was such a thing even possible? His track record convinced me, and I was determined to try, but I was filled with fear: I really needed my knee to be better. I could not imagine going through life as crippled as I was then.

In my moments of fear I felt tremendous shame as well. Shame of what, I did not know, but I felt I had failed in some terrible way, and I wanted to hide it, so I also felt defensive and secretive about the pain and the disability. I decided to move back to Ithaca, have my surgery, and then stay in Ithaca with Curtis and my parents. I sat in the office of the senior partner who had hired me and explained I had to go back to Ithaca to be cared for while I recovered from a big knee surgery. He was concerned and supportive, and asked when I'd be back. I'd done well and was welcome and wanted. Something in me shut down at the thought. Instead of going back to my job in Philadelphia, I started planning our wedding.

In Ithaca I found a job as the business manager of a magazine, *The International Wine Review*. While recovering from the knee surgery at home in Curt's and my sweet little half-house up on South Hill, sitting at the kitchen table with my leg in a brace and set up on a chair, I interviewed candidates for a job opening at the magazine. On that day I met Sarah Ploss, and we began our friendship. Sarah handed me her reference letters. Her confident voice and steady gaze made me trust her immediately. I hired her. Sarah, as clear-eyed then as she is today, watched me unravel over the next couple of years as I tried to suit Curtis's family, and in doing so, opened up my buried childhood issues and had my identity dissolve.

But in that season, all was wonderful. My knee was healing. We were preparing for our wedding, to be held in July at that Chesapeake Bay farm. Curtis, the scion of the family, was worthy of one of the biggest weddings of the social season. His mother presented us rapidly with her short list of the most critical one hundred people to invite, with hundreds more to follow. We looked at each other wide-eyed: We'd imagined a wedding of about one hundred people in total. With much compromising and conversation all around, we ended up with about 150 people, dancing under a tent. I wore a dress of my own design, with my mother's and grandmother's lace mantilla over my short dark hair. My brothers were there, Tom with his face in a funny expression halfway between happy and proud. Curt's cousin sailed down in his gorgeous yacht and anchored just off the site of the wedding, all flags flying up and down the halyards. The next day, Curt and I left for a honeymoon at the Bitter End in the British Virgin Islands, where we sailed and snorkeled every day.

Back home, I wrote hundreds of thank-you letters for the wedding gifts and well-wishes, many to people whose names I knew from newspapers and history. Curtis wrote his thesis and then won a prestigious congressional fellowship and we prepared to move to Washington, D.C. We were moving from the town with my parents, the town I had grown up in, with skinny-dipping in the creeks and hiking in the woods, to the town with his parents—where his father, Buff, was soon to be sworn in as an assistant secretary of state—and his cousins who had helped finance the Kennedy Center. I had declined to change my name and I still believed that he and I were both pursuing our own goals.

We rented a brick house just off Massachusetts Avenue outside the D.C. line. Curt's grandmother came to inspect it. She said it was respectable enough for us, but the red salvias in the flower borders had to go: they belonged only at train stations. Once her pronouncement was made, the family had a sigh of relief and Curt and I were allowed to go ahead and move in. I eventually landed a job on the business side of *U.S. News & World Report,* and I commenced riding the bus downtown each day. Curt became the science advisor for Senator Bill Bradley. We went to holiday dinners at the Cosmos Club, and swimming and hockey in season at the Chevy Chase Club. I never quite had the right clothes.

Sarah Ploss came down to visit. Janet took Curt, Sarah, and me to the Chevy Chase Club for drinks and dinner. Sarah's bright friendliness was deflected by Janet. I felt myself charmed and wakened by the joy and authenticity of Sarah and the echoes of the life in Ithaca she brought with her. But I felt the disapproval of Janet. Sarah had a few silent pauses over my newly conservative clothing and the strict table manners that told me she thought I'd sold my soul. My friend, the barometer. After that visit, our friendship was left on the frozen tundra for a few years, until I found myself again.

Down at the Eastern Shore on the weekends I felt a little more in my element. I could school Janet's Thoroughbred, a mare off the racetrack who had not really learned how to trot; she just walked or cantered. Or galloped. I could help prune the trees. I was learning to sail, and I was good at canoeing, and a demon at croquet. I did often say the wrong thing, though, and my sense of humor continued to be wrong: too sharp, too literary, too loud. The voice of the teacher from AP English would ring in my ears again: wrong poem, wrong choice. Up in the city, we traded social events with the family. Curt and I had a pre-Thanksgiving dinner party at our house for the whole family, and Buff and I traded some inappropriate jokes that had us both weeping with laughter over a big Toblerone bar.

AND THEN ONE NIGHT IN THE RESPECTABLE BRICK HOUSE IN WASHington, the phone on the bedside table rang, waking us. My mother was on the other end, telling me that my brother Tom had been killed by a drunk driver. I was twenty-four, and my life as I knew it was over.

I felt as if the world was getting louder and then softer, expanding and contracting around my head in waves. I was trying to listen to what my mother was saying. She was using her most soothing voice, the voice she used when a child or an animal needed to be calmed. What did we need to do? What happened next? Those questions felt impossible. This present could not be real, could not lead to a next step. This present immediately had to vanish into the hell it had come from and return us to what was right and normal. That feeling stayed with me for years.

The next day we drove to Ithaca. I think Curt was with me, but at that moment, it might as well have been just me, my parents, and my brother Jim, alone in howling darkness. Up to that point I had had the illusion that my family was tightly linked, like we were all holding hands in a circle. Now, with Tom cut out, our hands flew apart and our circle broke open and we no longer faced each other. My father was so stricken that he could not speak of Tom and for many years turned away when Tom was mentioned. I do not know what Jim's internal process was; mainly I experienced his criticism that how I was grieving and what I was doing wasn't right. And Mom had been gutted. She cried all the time. She decided there would be no service, and when we received Tom's ashes from California, she took them in secret up into the woods on the hill behind the house and spread them by herself. We were cut out; we no longer existed for her either. She stopped eating. I pleaded with her to eat, and made her favorite dishes. Jim and Dad only told me they did not want to "interfere."

Tom had been working as a postdoctoral fellow at Stanford, and then at Berkeley. He was a rising star in neurophysiology, and like so many people in that field in the 1980s, he was studying fruit fly genetics, working to understand the complex relationships between genetics and neurology. He was strong but slight, with a ready smile, curly dark hair cut short, and the soft look of a humble connection with you in his eyes. Growing up, he had been in the shadow of the brilliance of Jim, two years older, and I'd been in both their shadows, eight years behind Tom and also a girl. When I was a child I wanted to do everything with them. They taught me to walk everywhere in my bare feet. When I was very little and wondered where they went during the day (to school), they told me they were fighting dinosaurs on Mars. They told me it snowed in all different colors but I happened only to have seen white so far.

In high school, when I might have wanted to be out with my boyfriend, I canceled everything to be home every moment they were there over the holidays. There was no higher imperative than to spend all my time with them. Ever since I was ten they had both been away, mostly—college, and then graduate school. Without them, the house was cold and silent; the family with them gone was unrecognizable. But when they were home on vacation, anything was possible. The lights were on, the heat was on, there was cooking and food in the fridge, there were ideas and laughter.

Tom brought home things he thought I should know about. He brought me the album *The Roches* after he heard it at Cornell. He and I collected Beatles albums from the secondhand record store in Collegetown, working our way toward a complete collection. He introduced me to the right kinds of hiking boots, and how to play Frisbee, and the virtues of European chocolate. When he was in graduate school, and I was in college, I went to Madison, Wisconsin, to stay with him and his partner, Andie, for a week or two. I saw what it was to have an affectionate relationship, and a charming apartment, and jobs you wanted to go to each day. Andie, my dear heart to this day, and I sang Motown while we drove around looking at garage sales. This felt like my inevitable, desirable adulthood, just waiting for me, just ahead.

But that day in 1989, he and Andie had taken her visiting brother to Yosemite for the day. Another driver began driving toward him, playing "chicken," veering into Tom's lane to try to get him to veer out of it. As it was told to me at the time, Tom pulled off the road and stopped, but the other driver hit them anyway, killing Tom instantly and injuring Andie so catastrophically that she was hospitalized for six months. Her brother, in the back seat, suffered only a couple of broken bones.

The driver of the other car was drunk, driving and drinking all day in the park, we heard. He had his wife and children in the car with him. He was tried in federal court. I talked to someone from the court and asked if I could submit a personal impact statement, but she apologetically told me the date for those submissions was over. She had gotten one from my mother, but the court was not aware Tom had had any siblings. The driver received a sentence of five years in jail. I do not know his name or what became of him.

Mom was living in her own private hell and we were not invited. She had also had a deeply beloved older brother who was killed serving in the navy during World War II, while Mom was in college. Mom seemed to have had some kind of breakdown when he died. She couldn't remember much about that time. She went to four colleges, one for each year of her four-year degree. Now, with the death of Tom, she lost her will to live.

I still have an aversion to telephones. I hate the sound of a phone ringing and I will do almost anything to avoid making or taking a call. Each time I get into a car, or when I know someone I love is driving somewhere, I have to remind myself: this is a normal risk we all choose to take. I was twenty-four when Tom died, and he was thirty-two.

CURT HAD REACHED THE END OF THE TIME OF HIS FELLOWSHIP IN Washington, and needed to find a next job. To my surprise, he started looking for postdoctoral positions all over the country. I had some-how thought we were settled in Washington, D.C., for a while. Then, one evening, as we ate dinner with his parents in the beautiful dining room of their Spring Valley, Virginia, house, he announced his frontrunner job option was in a small college town in the south. A little shock ran through me.

Wait, I said, what would I do there? Everyone turned to look at me in silence.

Later, Janet pulled me aside. "Your job is to be with Curtis," she told me. "You don't get an opinion on where his job is or where you move. Your job is to support him." I heard the anger in her stern, flat voice, and I felt my failure, and my inability, to accept and appreciate what life in this family meant.

We moved, instead, to Annapolis, where Curt got a job with the Chesapeake Bay Foundation. We rented a little wooden house a few blocks behind the Naval Academy stadium, almost on the shore of Weems Creek. This felt more like home. We had a long backyard that sloped into woods, and a vegetable garden. His parents and I bought Curt a solo canoe for his birthday, and some days he commuted to work in it. I started a consulting business writing business plans for young companies, and we started trying to have a baby.

In relishing the woods and the creek and the vegetable garden, in all this reaching back to find the pieces of my childhood that had beauty, I was also reaching for animals. My knees were too bad to allow me to ride horses at that time, so I turned to dog training. I had met some new friends who raised sheep and trained sheepdogs, and I apprenticed myself to the generous, kind Barbara Starkey, and became her farm assistant a day or so per week. And then I got my first border collie, Kline. Choosing the dog reminded me of Tom, because he and I used to make lists of our favorite dog breeds, and discuss what the best combinations would be for our eventual housefuls. We made lists of dog names. Kline joined Curt's Siberian husky, Natasha, in our household, and we started obedience classes together, and life felt right, and sure, even as my heart was still broken over Tom's loss and I had more and more nightmares.

One day about six months after Tom died Curt saw me with tears on my cheeks, and I recall him saying, in surprise, "I thought you would have forgotten about all that by now, or gotten over it." Responses flashed through my mind. I will never be over it, I thought. And who in the world in six months is over a terrible personal loss? Where was the foothold to start a conversation about this comment? Could it be that Curt's family had treated any deaths that had occurred during his lifetime without emotional reaction? But I couldn't reach that far from my own grief into a dispassionate analysis of his question. Curt had asked not with malice or resentment but instead a bit of surprise and wonder. I realized he had no emotional bridge to where I was right then. I was alone. Our lives went on as they were going.

At last I was pregnant. I went to the doctor's to be checked because I didn't know whether to trust the drugstore tests, and I happily gave Curt the doctor's note of affirmation that evening. We were thrilled! His parents were thrilled! And on a visit to Ithaca with my parents, I sat them down and excitedly announced my surprise: we were going to have a baby! They looked at me steadily but a little blankly. With an internal shudder of embarrassment, and a feeling of floating into an empty space with no anchors, I realized that they had not been waiting for this moment. Having grandchildren was just not on their life list of joys to come. They were pleased, nothing more.

But the joy of the pregnancy was strong for us. We talked about names. We thought, if it's a girl, she'll have my last name, and a boy, he'll have Curt's. Curt's parents did not approve. My parents did not care. Curt's lovely cousin gave me a baby shower. What fun to open all the gifts! And our new, to-be-lifelong friends Ilene and Philip from a couple of blocks away gave us their own old crib.

By the time I was six months along and Curt's family had taken us on a chartered sailboat trip out of Saint Lucia, I was no longer nauseated, and could snorkel and bodysurf along with everyone else. Curt's beautiful, socially graced, and wonderfully welcoming and loving sisters were along, of course, and they knew exactly how to make everything smooth. I did not. Janet criticized the weight I had gained. And I needed to drink a lot of milk, and bought shelf-stable milk to keep in our cabin; she wanted it with the rest of the food, and when I insisted I be allowed to keep my own milk, she was angry, and I felt selfish and stubborn and wrong.

I kept having dreams that I was having a little boy who looked like Tom. And then, the day before giving birth, I had a dream that I had a little girl. I was absolutely sure I was having a girl; so easy to explain away the boy dreams as missing my brother. So when, the next day in the hospital, the nurse said, "He's a boy!" I was so surprised that I could only say, "What?"

But this gorgeous, healthy, big baby boy looked so much like the baby pictures of Tom. We named him Turner, which was Tom's middle name, and Kolbe, a last name in my mother's family, and Bohlen, his father's last name. Turner Kolbe Bohlen. So beloved from the very moment we met him. What a miracle a baby is.

WE LIVED IN THAT SWEET HOUSE IN ANNAPOLIS FOR A COUPLE OF years, and then Curtis needed to change jobs again. I asked him: "Could there be something you are struggling with that is not the job? Is it really the job that makes you unhappy?" But just as I was unable yet to start owning and shaping my own pain, so perhaps he was, too. The conversation was short as this paragraph. He took a new job at a biological station down on the Chesapeake Bay. We needed to move far south from Annapolis, down to southern St. Mary's County, along the western side of the Chesapeake Bay. We found a little place with some land—eighteen acres—so we could raise sheep of our own and continue training our border collies to herd.

We set about converting the property from an old tobacco farm into a new sheep farm. We built a seven-strand electric fence around a ten-acre pasture. We had hand tools, but no tractor and no big truck, and no way to plant or hay. A dirt lane ran along the far side of the sheep pasture, through a hedgerow, and along fields to our neighbor's house. One day soon after we moved in Curt and I drove over there. We stopped in the dirt area in front of the house, called out hello, and waited by the car. We'd already learned the country practice of waiting out by the vehicle and not coming up and knocking on the door, which was much too invasive. No sound came from the house. No door opened, and no curtain moved. Eventually, we got back in our car and drove home.

A few hours later, a little contingent came walking down the lane to our house. We met Nancy Zimmerman, and her husband, and several of their ten children. Living around us in southern Maryland were members of several different sub-denominations of both the Amish and Mennonite churches. Nancy and her family were Mennonites. They used only horses for transport, and eschewed even buttons as vanity, closing their long dresses with pins. Her oldest daughter was married and gone, but her littlest boy, adorable and mischievous, was Turner's age. We exchanged greetings, and names, and a little history, and I was glad I was wearing pants at least instead of shorts.

Nancy let us know the name and address of the Mennonite man who had been farming the land we purchased. Later, Curt and I went to meet him and we made an agreement by which he would hay our fields in return for keeping half the harvest. This was enough hay to feed our sheep all winter, along with the neighbor's team of working Belgians. These lovely horses pulled a hay cutter in season, and then a bailer. We'd all walk alongside through the hot summer day, dust motes from the hay rising up in the warm fragrant air, picking up bales and throwing them up onto the flat wagon. We'd open the barn doors wide on both sides of the barn, and the horses would pull the wagon straight in and stop in the shade so we could unload.

Turner went to play with the Zimmermans for a couple of hours in the mornings while I wrote business plans. Some afternoons, I took Nancy and a couple of her kids with me and Turner to the grocery store in our SUV. She was allowed to ride in a car, just not to own or drive one, and this was much easier than shopping by horse cart. For a special treat, we'd eat at McDonald's. Once, I drove us all up to Washington and we went to the zoo. Against the odds, we had really become friends, bonded by our children and common experiences as women. We talked about that sometimes, which freedoms were allowed us, and which were not. We both chafed. Nancy, with her long-sleeved, long-skirted dresses and bonnets, chafed sometimes because she could not spare the cows and horses their long, painful, neglected labors; her husband would never call a vet.

Other afternoons, Turner and I would take care of our growing herd of Border Leicester Cheviot–cross sheep. I bought a small trailer and built up hog panel sides and a top. Covered with a tarp, this trailer was good for moving sheep from farm to farm. I bought ewes from various friends, and then a lovely young ram from a Border Leicester breeder. This ram had a naturally brown fleece, perfect for my breeding plans to make fleece for hand-spinners, and freezer lamb for epicures. I named the ram Abraham and the first purebred ewe Sara. Curt and I went to shearing school, and we sheared our own sheep and a number of close-by small flocks as well. At lambing time, tiny Turner would go into the jugs, the little pens each ewe had for herself and her baby for a week, and pick up the lambs for me to check on. The ewes and lambs were not alarmed by him.

Every evening when the temperature cooled, Curt and I would take our border collies down to the pasture for training. We'd gone to clinics, and trained with our friends, and over the years learned enough to be useful at this. Turner would play in the clover or ride in a backpack. We'd have the novice dogs wait while we walked away from them toward the sheep, and then we'd send the dogs one way or another on their sweeping arc around to the other side of the herd. "Way to me!" we'd call, or "Go bye!" More advanced dogs were learning their whistles, how to drive sheep, how to split herds and keep the halves apart. Mennonite and Amish carts would line up along the road to watch; most of their passengers had never seen working dogs before. These fast, intense, soaringly intelligent dogs were the most beautiful things I had ever seen. They were pure intention, pure determination, pure speed, like liquid missiles as they streaked away, only a single job in mind.

Most weekends we were off to sheepdog trials. We competed throughout the mid-Atlantic and sometimes up into northern Pennsylvania and even New England. One long weekend trial at a big sheep farm, my friends and I were sitting in folding chairs and watching the trial, when a "Big Hat" friend walked up. The Big Hats are the top professional handlers and trainers, the people who breed lines of working dogs, and compete at the end of the year in the North American finals, and who teach clinics to help the rest of us. This Big Hat, Beverley, told us that one of the puppies from the last litter of one of her all-time-great dogs had been purchased by the man who owned this farm and was hosting the trial. We were a bit aghast; he had a reputation for being cruel to his dogs. "Right," said Beverley. "He promised me this pup would live in the house with the family and have a kind and connected upbringing. But she's locked out in a kennel with her nose swollen up like a cartoon dog from pressing it through the wire mesh and begging for attention. If you go up to her she flinches like she's been beaten. I need someone to buy her from him and take her home. Would any of you four want her?"

Oh, I wanted her. But to part with her, the owner wanted $500, much more money than I had. But my friends lent me the money, and I bought her. She was not yet a year old, still so slim and bendable she seemed almost like a canine pipe cleaner. She was a bright, foxy red with a big white blaze and collar. I changed her name from Fly to Tess.

Curt had not thought we needed another dog—at that moment I think we had five or six—and he was exactly right, we didn't. We'd been taking in rescue border collies, training them, and placing them on farms. But he fell in love with her just as I did. At home, her abuse was obvious. Once, when she was under the kitchen table, I realized I was standing on her paw; instead of yelping or flinching, she simply lay down and went limp. She dissociated. Young border collies have the energy of nuclear reactors, and they are as responsive and connected as toddlers. When you lie on the ground, a puppy will generally rush to you and begin licking your face, or otherwise interacting. When I lay on the ground, Tess would run right past me like I did not exist. She had learned to exist without people.

And then one day I was working in my little home office, and I felt a touch on my leg. I looked down. Tess had crept in silently, and was sitting by my chair with her chin placed firmly on my thigh, and her eyes rolled up to meet mine. She'd made a decision to try.

Of all the dogs in my life, Tess was *the* dog. She and I were mentally connected. But she was more than a dog; we called her a higher being. She had a complete sense of what it was to be self-contained, to be responsible, to be polite. She was fully formed and self-determined. Once I offered her a piece of bacon as a reward for learning a little bit of obedience needed to prepare for herding, and she shut her eyes and turned her face away. I had insulted her completely. We were either willing partners working together, or we were nothing, her gesture said. Tess adored and watched over Turner, and over all the lambs. She learned everything I tried to teach her about sheep in a flash; she just needed to know the intent, and then she was on it.

One day when Tess was about a year and a half old, I asked her to move a big militant ewe and her two new lambs from one barn to the next. Tess crept up toward the ewe—working border collies crouch and move like predators, with a strong gaze, and that is how they intimidate herd animals into moving—and the ewe with her lambs at heel began to move away. A moment later, the ewe was struck with a doubt and thought this dog might be a threat to her babies, and so turned around and stamped her foot at Tess as a warning and perhaps the preface to a fight. As the ewe turned toward her, Tess slowly and smoothly averted her gaze to the side, without moving her body, just releasing the predatory pressure of her gaze on the ewe. The ewe, almost visibly confused by the sudden vanishing of the threat, turned around again and started moving in the right direction. Such was the delicacy and cognition of this amazing dog.

Turner was with me for all of this. We'd care for our eight beds of vegetables, and we'd undertake special projects. One was building a swing. A huge oak right next to our house had been killed by a deafening lightning strike during a summer thunderstorm (the lightbulb in the kitchen ceiling had burst, and I suddenly had four little boys from Turner's playtime in my lap, asking to go home). Curt had split it by hand into boards, for the pleasure of learning how. Turner and I picked a good one, planed it smooth, hand-drilled holes through it, and hung it from a high branch on the cherry tree by the house.

Daytimes with little Turner, caring for the sheep, visiting neighbors, were busy and challenging. But evenings with Curtis became harder. Curt and I went on fewer canoe trips, fewer sheepdog trials, together. He didn't ask about my day when he came home. I began to feel lonelier when we were together than when we were apart. As my loneliness became overwhelming, I began to daydream about escape.

And one day when Turner was four and tucked in bed asleep after dinner, Curt said he was going to leave this job and move to New England for a job at a college. I was standing in our little living room, where I'd had a neighbor build a window seat for reading, with bookshelves below full of Turner's books. Where we'd installed our woodstove for that comforting, penetrating winter warmth. I felt a wash of coldness sweep down my body followed by the adrenaline of panic. I looked out the windows to the barn and the pasture and the sheep, grazing contentedly, each one known, every ewe named. How could I leave the sheep and the dogs, and the farm whose grass and earth and trees I knew intensely? How could I allow Turner to lose all he had on this farm? My first words were, unthinking and true, "I can't move right away, I have to stay longer. I have to stay here." Curt immediately agreed: "Okay, I can go up and start the fall semester, and you and Turner can move up for the winter holidays." No, I thought, not enough . . . but then, at once, it came to me: I could leave Curtis.

A few weeks later, I was sure, and I told him. His desolate tears, and my own, felt like death. Curt and I were well and truly married. No matter the loneliness and the coldness that had increased, we were grown into each other, and leaving him felt like my body was being rent in two, disemboweled.

So in the middle of that winter I fled like a little wild beast. I left my husband of a decade and our house covered with clean white snow, and then I was on my own. When I was not crushed by guilt, I was ecstatic, floating, breathing the wine-sweet air of winter in my joy. I wanted to share that feeling of hope, the new energy and possibility I had, with Turner, to show him there was still love and joy in the midst of mourning. I moved into my own old farmhouse, rented on a frantic morning and inhabited that speeding afternoon. I was working three part-time jobs, trying to pay half the mortgage on our old house as well as my new rent, and I was happy as I hadn't been since I was twenty.

Imagine the strength of a freedom that could begin to balance the loss of that beautiful farm and flock, and the blood, sweat, and dedication I had put into building them, and even, to begin to dissolve the agony of ending a decade of committed love. I put what furniture I could afford into a bedroom just for Turner, who would share his time between the farm and Curt, and me. On nights when Turner was with Curtis I was alone with Tess, while I listened to new kinds of music and ate potatoes for dinner.

JAMES AND I HAD MET THE PREVIOUS FALL AT ST. MARY'S COLLEGE of Maryland, where I had gotten a job lecturing in mathematics and he was a professor. I felt I had written the last fresh word for a business plan that I had in me, and I'd found this job on the strength of my MIT master's. The chairman of the math department asked James to sit in on my classes and see how I was doing. A tall man with a relaxed and self-contained way of moving, he sat quiet and straight in a chair among the students. His bright blue eyes, a contrast to his dark hair, looked at me steadily and piercingly. At the end of class he came and said some complimentary words to me, words I can't remember anymore, the words that were the beginning.

We began a friendship. James listened with patience and seemingly without judgment to my explanations of where I was in my life. He wouldn't consider any deeper involvement until my separation was entirely clear and completed, and so it was with little expectation that I contacted him in the spring to explain I had moved out before the holidays, and was settled nearby. On our second date I cooked him dinner, and he taught me some contra dance steps in my kitchen before we went off to the dance. Those early weeks I discovered a new feeling: I would wake at five in the morning, heart beating fast, thinking of him. We'd go to lunch and I couldn't eat. I missed him in the evenings when we didn't go out together, and I rushed to work in the mornings to see him again.

My friends and family were not at all sure I was doing the right thing. Several of my friends questioned my being with James. Suzanne said she thought it was crucial for someone to live alone, to learn to live with themselves, before they could form a lasting relationship. Sharon said it was good to live alone and also that it was too bad someone as good as James was my first relationship after Curtis, thinking the first relationship couldn't last. I lost some friendships over the speed with which we started this relationship. My mother was kind but distant with James, but my father wouldn't look at him or speak to him. Kind, gentle James. But whether I was being foolish or not, I was helplessly in love.

One morning, going through boxes at the rented farmhouse, I found an unopened box of paints Curtis had given me for Christmas last year; he must have bought them right before I left him. We didn't spend that Christmas together. By Christmas I was gone. Then, in another box, I found the first book Curtis ever gave me. He was not greatly demonstrative in other ways, but I could infer how he felt about me sometimes by actions he took, like buying the paints. I had been struck at Christmas by the kindness and focus of his presents, better than they had been in years, nice presents, well chosen. He'd dropped them off for me rather than throwing them away.

I suddenly felt poisonous, like I ruined and poisoned all the things around me. This kind of plunge from happiness and the normal activities of life into despair was happening more and more. Suddenly I'd feel unbearably despairing and usually have no idea why, what the trigger had been. I was scared by the depth and complexity of these emotions. One thing, like reading someone's misspelled word, would cause me to feel panic and sadness almost to the point that I couldn't keep normal conversation going.

So what was wrong with me? I'd had some intense bouts of depression before in my life, but they'd been gone for some time. I thought I might have to face the depression again, someday, perhaps have some therapy and find out what it had all been about. Then, inevitably, over the next summer while James and I walked in the woods and went contra dancing, the sickening depression moved in again. Instead of waking with a bit of cheer for the new day, I'd wake feeling disoriented, still lost in the sensations of confusing and frightening dreams. In the afternoons some little event, an unexpected phone call, a nasty glance from another driver on the road, would sink my stomach and I would feel hopeless. The hopelessness enlarged like a flood of icy water to take over all my feelings, and then I felt completely lost, lost to hope and happiness, which are meaningless in this state, but moreover, lost to myself: I didn't know what I wanted or even what was wrong. I experienced a deathlike state, detached, helpless. This state came on in a matter of moments, and then after an hour or a few hours or a day, it disappeared, and I never knew why. I was in a deadly oscillation between relative stability and complete paralysis and drowning depression, an oscillation that could happen more than once in a single day, or descend for days on end. All this despite the wonderful life I then had with James, the new hope and independence I had found in my job. Most mornings that summer I awakened, looking out the second-story windows of the old farmhouse I had rented, flattened with unremitting, helpless fear and dread.

My life was busy and also, ironically, productive, despite my depression. The very constancy of my work and attention to things that fascinated me allowed me to block out the greater fears, and ultimately, to block out the memories and to keep moving. I worked hard and constantly through all the pain and depression and anxiety. I taught three math classes per semester, and I wrote poetry. If I had stopped reading and writing and working I certainly wouldn't have been able to withstand the pain of the depression.

But I knew I was drowning as surely as if I had given up and gone to bed; I was just drowning more slowly. I didn't know how therapy would help me, or even what it was, really, having had no previous experience with it either personally or through others in my family. Therapy wasn't something my family did, but it was the only thing I could think of that might help me. When you're depressed you're meant to have therapy, aren't you? I called the office of a therapist my doctor recommended, and by great good luck made an appointment with that young Jungian analyst in training, Mary.

Chapter 3

BEING RELENTLESS

What was my childhood like? I do not know how to answer that question. I can tell the story of a happy childhood: I grew up in an upper-middle-class family in Ithaca, New York. My parents married shortly after the end of World War II, and they stayed married for over sixty years, until their deaths in 2007 and 2008, in their early eighties. I was the youngest of three children and my big brothers, Tom and Jim, often doted on me.

We had a pond filled with frogs that sang me to sleep in the summer, and we had dogs and rabbits and cats and mice and guinea pigs. But our household was eccentric. We had a rope swing in our cathedral-ceilinged living room. My mother varnished *National Geographic* maps onto the floor, so we would learn to crawl while looking at the countries of the world. We were each largely left to make up our own minds about what to do with our lives. My mother said that a child should be able to make their own decisions by the time they were ten. We were left to play as we liked, to climb trees and make forts and explore Palmer Woods. We were unsupervised. Someone repeatedly raped me.

The fact of those assaults was with me always, like a gray veil floating before and behind me, a family secret that was kept from almost all of the family. Unspoken. Much safer to leave it unspoken, and to constantly press to be out in front of it, not falling under the crushing depression or paralyzed by the overwhelming terror of those memories. Much safer to pretend so hard that it never happened that I could make a life that mainly did not include any part of it.

We were not allowed to have nervous breakdowns or be depressed or get counseling. People in our family did not do that, said my mother. My mother herself was deeply depressed for much of my life. Years later, when I brought it up to her, she did not remember that she had a bottle of sleeping pills that she showed me, when I was in elementary school. Nestled among the fragrant silk scarves of her proper Philadelphia upbringing was this bottle of pills she was saving up to kill herself with when she needed to, as she explained to me at the time. She did not remember when I was seven or eight that my anus was so sore one morning that I leaned over against a chair, weeping. She told me she was busy and could help me a little later. She did not remember that I had a vaginal infection before I knew I had a vagina.

When I was ten, I was diagnosed with scoliosis, curvature of the spine. My doctor decided I needed to start wearing a Milwaukee back brace that started with a leather girdle around my waist and hips, and rose up via three metal bars, two in the back and one straight up the front, to end in a metal collar just under my chin. Leather pads clamped to me by the stiff leather girdle pressed the curve in my spine back toward the midline, and the collar kept my head up and my spine stretched. I would wear that brace until age seventeen.

I had been taking horseback-riding lessons since I was seven. From the time I could hold a pencil I was drawing horses. I read every book about horses I could find, starting with the read-aloud books in the children's room at the public library and continuing with the shelf of animal books in my elementary school library—I can picture it so clearly, which aisle, which shelves, my search for a new one I hadn't read. My only sweet dreams were about horses. I asked for lessons for years before my mother finally let me go to riding summer camp, which remains in my memory as a kind of sun-kissed, horse-scented heaven populated with the ponies named Barney and Bunty that I most often rode.

At ten, when I got my brace, my parents decided to buy me a pony of my own as a kind of recompense. The doctor did not want me to ride, because of the vertical compression of the spine, but my parents so wonderfully understood that if I was forced to give up riding they might as well kill me right on the spot. I had already had to give up gymnastics and swimming because I was only allowed out of the brace for an hour a day. The sexual assaults ended at about that same age, and I bonded more and more closely with my Shetland sheepdog, Sunshine, and I spent as many hours as I could at the horse barn: the warm scents of hay and horse, the sliding feel of horsehair under my fingers as I held my hand against the pony's neck under their mane, the speed of trotting, cantering, jumping, running away.

I was taking lessons with Gary Duffy at the GlanNant Welsh pony farm, owned by Welsh pony expert Mollie Butler. Welsh ponies can be as refined and as talented as Thoroughbred jumpers. No shaggy coarse backyard beasts, these ponies held their alert ears above dished foreheads and bright, knowing eyes. To get to the indoor riding arena, you walked through an aisle with a tack room on the left, and then stalls either side, perhaps six of them. Coming through the main doors, I was always met by the rustle of pony feet in wood shaving bedding, and the occasional nicker, faces looking out over the half doors, and the sound of the instructor's voice from the arena, and of hoofbeats. The rich smells of horse and hay and sweet feed filled the air. I was home.

I began my lessons riding a calm older mare, GlanNant Lyric, and her roan cousin, GlanNant Rhyme. Lyric felt tall; I was a little girl, and Lyric was a medium pony. Her face was open and peaceful, and her thick white winter coat especially soft. On Lyric I learned to post properly to the trot, and I learned how to canter. I was not a courageous rider, but I was determined to keep trying. Even the most lovely of these Welsh ponies could shy like a jackrabbit—one excuse to be frightened, and the pony had dropped a foot in height and shot sideways three yards in a split second—and riding them did develop a good seat. I rode little bay GlanNant Primrose Lane with her intensely pricked ears in my first horse show, in the big echoing arena at Cornell University. I learned later that she had won national pleasure pony of the year, and national reserve champion pony hunter, huge prizes in horse showing. She was kindly slumming it with me, that day when we won some ribbons in the show division known as short stirrup.

After lessons, as the early dark of winter came, I'd stay and help with chores. We went to each of the stalls in the lesson barn, broke the ice in the water buckets and refilled them, and gave each murmuring, stamping, nickering pony his or her share of scoops of sweet feed or pellets, and flakes of hay. The hay was sweetly fragrant even in the coldest of freezing temperatures, and the sweet feed smelled good enough for me to want to eat it. The heat radiating from the ponies' beautiful round bodies warmed up the cold barns. We'd go next to the broodmare barn, and feed the mares and their half-grown babies from the last year's crop of foals. Finally, the stallion barn. To keep the feisty stallions company, Mollie kept an ill-tempered billy goat. We had to negotiate getting into the barn without letting the goat out, and without being in the barn with the goat. He had to go into his own stall. If he escaped, as he sometimes did, he'd chase us back to the lesson barn and we'd jump into an empty stall and slam the door behind us while he charged, horns lowered.

When I got my back brace, I still felt I belonged in this world. Along with Sunshine, riding was what I loved best. We began to look for a pony for me. Gary brought in ponies on spec, to try out. One I really lost my heart to was GlanNant Catherine, originally bred on this farm but sold, and now available again. She was a lovely little dappled gray pony with bright, lustrous eyes and a hint of defiance. She was quite overweight, and I began to think that was because she was impossible to exercise. During the day, I wrote "Catherine" all over my school papers, daydreaming, and every afternoon she beat me up in the riding ring or in the field. She shied or bucked me off. She refused fences and I couldn't force her. She ran away with me and I can still feel the bit, iron in her teeth, and nothing my young body could do from the saddle and with the ineffective reins meant anything to her. With the feeling of losing a love, and with the feeling that I had failed, we sent her back.

And then we found Farnley Drumbelle. Drummy was a medium-size bay mare, already fully trained ("made") when I met her. She was an expert at bringing along young riders. She wasn't the most beautiful pony—her back was a little long, and her face not as stylishly dished as some Welsh ponies' were—but she was the wisest, steadiest, best friend I could have. She taught me to relax, and she taught me to jump, and she taught me how to show, and as I improved, she allowed herself indulgences, like a shy, or a buck, or a bad mood. Eventually she made me good enough that I could turn to training new young horses and bringing them along. She gave me my handholds along the slippery cliff I was walking along on my way to escape to college.

Then, as does every little girl and boy rider, I outgrew my pony. As was custom at the stable, we sold Drummy to a younger child. Then I had a choice to make: pursue the junior hunter and equitation competitions with the hope of ending up in the Medal-Maclay Finals in

Madison Square Garden, one of the paths to the Olympic team, or stop showing and instead train young horses. I had had a lot of success in horse shows—my room at home was doubly ringed with show ribbons. I loved showing, and I also hated it. In the beginning I would cry and even vomit from nerves before going out in the ring. And I saw how winning became more important than the relationship with the horse, as expertise progressed.

So, I finally opted to stay home at the barn and train. We bought an ethereally beautiful, nervy Thoroughbred-Welsh-cross filly and I named her Imagine That, after John Lennon's song "Imagine." She and I struggled but we also understood each other and trusted each other eventually. She was exceptional. Beautiful, tingling with life, she loved to jump, though she also fought with me every step of the way. In the end, I trained her up and then sold her to go onto the show circuit, as I had promised my parents. She won a huge national prize, and another, and then she kind of vanished from the scene. I don't know what happened to her, but I worry that someone fought with her the wrong way, soured her, or hurt her, or broke her spirit. I heard she'd started refusing fences. Even thinking of that mare refusing, that mare who had gleefully jumped over the five-foot fence of her pasture, galloped around the cornfield, and jumped back in, just for the joy of it, broke my heart. Over the next twenty years I would occasionally dream I went back into the barn and she was there waiting, nickering for me when she heard my footstep, as she used to do. I swore that if I ever had another horse I would care for them until they died. But after we sold that mare, I never did return to the barn. Then, I was just home with Mom and Dad.

MY MOTHER HAD A PIT OF INSANITY THAT SHE SPENT A LOT OF ENERGY trying not to fall into, but sometimes, she slipped. She had nightmares in which we, her family, took her to an institution for the mentally disturbed, dropped her off behind the locked gates, and drove away. This never happened, but she was, I think, often taking her own internal temperature: *How far off normal am I now?*

One day, when I was in my late twenties, Mom and I were sitting on the porch of the Maryland farmhouse James and I had rented, when she told me she wished we were closer.

"We never, ever bonded," she said. "Not from the moment you were born. You didn't bond with me. Maybe because I smelled of cigarettes. You may have felt how much anger is in me, too. Anger directed at Dad."

"Yeah, I could feel that," I replied.

"I never really had you," she said. "I never *had* you." Her gaze was turning inward. As I was puzzling through what those words might mean, they recalled for me her repeated story, one I'd known my whole conscious life, that she had burst into tears when I was born, because she had never wanted a girl. She didn't know how to bring up a girl.

"Was I meant to be a boy?" I asked her now.

She didn't answer. "I would have had a fourth child, you know," she said instead, "if I could have been assured that it would be a girl. I just wouldn't risk having a boy after a girl, because of the girl-dominant-to-a-boy problem."

"What problem is that?" I asked.

Her gaze had turned further inward, and her pronouns became confused. She'd gone into a frightening place by herself, and she was suddenly talking about herself. "Where a girl, being older, learns to boss around and take care of a boy, then I learn to be superior to him, and that makes it really hard for me to have a successful marriage. What chance does she have of finding a man who will let her be superior to him? I see so many marriages with hard, superior women and it takes all the manliness out of the man."

Some years later, Mom said she had one main thing she hoped I might talk to her about, one big question she had had all her life, if I ever had an answer. I asked her what that one big question was. She said the question was why I had pulled away from her, refused to learn from her, stopped cuddling and running to her. I asked when that had happened. As she said "about age four or five," a chill ran through me. It hadn't been from birth . . . it had been from the age I was sexually abused. I made this connection immediately, though I do not think she ever did.

As I was growing up, Mom encouraged me to look pretty and take care of my hair and skin, but there was a line between attractive and slutty that I couldn't always anticipate. My red halter dress was sexy and gorgeous, in her view, but lip gloss, even Vaseline, was slutty and not to be allowed out of the house. Far from promiscuous, I had hardly had a first kiss before the end of high school. Yet once she told me that the bathing suit I was wearing—a Speedo one-piece—made me look like a whore. I tried out the word "whore" as having some relation to myself. And so to me my mother's reaction connected only to the female body, since her judgment could not refer to my actions.

Throughout middle school, high school, MIT, and beyond, I suffered from depression and anxiety. Every day when I lived with my parents I awoke to the alarm clock with a shock and an instant, devastating, sinking sense of despair the moment I reached consciousness. My mother described me as stubborn and determined to get my way, but I never felt I was pushing others aside, or that I was callous—I always felt vulnerable and lesser than those around me, and was simply determined to keep moving forward on my own road.

And even beyond those things and other things I do not know how to fit together into a whole story, a single story, are the things I do not remember. Or, I remember them for a while, and then I forget them again, my mind pushes them away. Years of my life are vague and dreamlike. I cannot recall the immediacy of the emotions I had at those times; I don't feel like I inhabited my body. I watched. I realized this first when I was in junior high school, when I looked back and couldn't remember all of my elementary school years.

And sometimes when I remembered those things or came close to them, when I was in my twenties, I was suddenly compelled to lie down and go to sleep. As if they knocked me unconscious. I have slept, helplessly and irresistibly, under a tree, when I revisited a forest where something bad had happened. And in the same way, when I felt the terrors of an event in my body, I have slept on the top of a huge red rock in the desert, and I have slept at the side of a field of soybeans.

Sometimes I read about people completely incapacitated by grief, undone. A beloved person dies, and the living wail, collapse, shriek. Their insides become visible to the world and their friends rush to support them. But in the moment of a great tragedy, that doesn't happen to me. Instead, I see the precipice rushing toward me—I stop short, I say no. The fear of losing control to that bottomless black ocean of grief is too great. If I fell off that precipice, I am afraid I will fall so far that I will die. So I do not fall.

But my dreams are another matter. From the earliest childhood I can remember, I had recurring dreams that were inhabited with the worst kind of terror, a fear so complete that I don't think more fear could be supported by the mind. In these dreams I would be overcome by a shuddering and trembling terror that began in my legs and moved upward, like a knife-edged quavering scream from the depths, but I was absolutely mute.

In my dreams, the source of the terror varied. Sometimes large men or beasts pursued me through a rolling, cartoonish landscape, and I tried to hide under a bush or behind a rock, but I was always found. Sometimes I hid from terrible monsters inside a house with walls full of cracks and open windows and doors that did not latch.

Sometimes wolves stood tall and silent, silhouetted against the nighttime windows of my bedroom, and my only hope for living was to lie absolutely completely still. Sometimes it was a fast, long creature like a little weasel that would dart out of the most unexpected places and race up my leg. Always I was facing some kind of annihilation worse than death.

The terror in my dreams is the strongest emotion I have ever felt.

IN COUNSELING, I TOLD MARY THESE STORIES FROM MY CHILDHOOD. I didn't really plan ahead what I was going to talk about, but stories and memories came up in my mind and lingered, and then I wanted to talk about them. I began to feel as if there was a little monitor in my mind, doling out these stories and bringing up memories as they were pertinent. I began to feel that the hidden parts of my mind were ready for therapy. I remember that one day in my early childhood there was a dogfight in our backyard and Mom ran out of the house with an ax to stop the dogs fighting—an ax! What was she going to do with that? I told Mary about the bottle of pills my mother was saving up to kill herself with.

Mary gently commented, "You know that's not normal, don't you?"

And I could only shrug. Sure, these moments were upsetting, but how could they explain the well of despair inside me? I also remembered Mom cheering me on in sports, taking pleasure in all my interests, allowing me to keep lots of pets, and lovingly tucking me into bed each night. I remember Dad teaching me the Latin names of plants in our woods, and picking berries with me, and taking me to museums.

I asked Mary if this was a normal depression, this rapid oscillation, this frenetic worry, this devastating frozen ocean of despair. She said no, normally you can't get out of bed, are sad all the time. She said only a part of me was depressed, just a part, but it was really, really depressed.

In the midst of all the active therapeutic work, James and I went to my parents' house for three days. There I had continuous formless nightmares all night, and intense fear of the dark and of the woods. I had not been afraid of the dark for years. Now I was afraid of dark rooms, and of walking outside the house in the dark. Every night was a morass of half wakefulness and these awful nightmares.

I went to another therapist, a psychologist, for an evaluation. After seven sessions and some tests, she said her assessment was that I had spent much of my childhood "in a state of terror." She diagnosed me with post-traumatic stress disorder. The trauma I was carrying was repeated sexual abuse when I was a small child. Into my late twenties, I had some of the physical manifestations typical of PTSD. The most common was the feeling I was being crushed and suffocated when I was lying down in bed. My heart would beat in fear, and my chest would feel tight, compressed, crushed. My eyes would be droopy and my mind would be drifting, but my body would be as tense as a board, every muscle stiff and anguished, the pressure of my back against the bed almost too much.

And that feeling of terror. Beyond any horror movie, beyond anything I've ever consciously experienced. A shriek so primal it might have rent the bright day into shreds and revealed the blackness behind it. I think an echo of that terror was brought back by Hans Bethe when he described mutual nuclear destruction, and with that echo of the terror, the depths of time and the length of the solar system seemed they might be enough for safety.

Over five years of therapy, taking step after step, asking question after question, my courage grew. I learned I could survive when the pit of depression threatened. One by one, the recurring nightmares I'd had my whole life vanished. My mornings began with pleasure instead of dread. The terror receded. In my childhood and early adulthood everything was frightening; going into a bank to make a transaction was frightening. Going into a library. Planning travel. Anything new, anything where I would be under the scrutiny of others. As the talk therapy continued, though, those fears simply dissolved away, bit by bit: sometimes a painless vanishing, and sometimes a painful revelation that took the fangs out of the fear, and sometimes a little recurrence, but then always an ebbing away. By the time I was in my mid-thirties, almost nothing was frightening anymore. The terror was gone, and forging forward in life was both possible and usually a pleasure. The transformation was so complete that I still find it unbelievable. I have not had one of my thirty recurrent nightmares now for over twenty-five years. The poet Natalie Diaz writes, *You are not the sum of your injuries.* But perhaps I am the sum of the injuries that I have overcome.

Janet, Curt's mother, invited me to visit her as she lay dying after a long battle with cancer. I was able to tell her how much I loved her and appreciated all she had done for me. And she told me, with generous forgiveness on the eve of her death, that I had been a wonderful daughter-in-law. And now James and Turner and Liz (Turner's partner) and I often have holiday dinners with Curt and Buff at Curt's sister Julie's house; she and her husband and daughters welcome us all in and we feel that warmth of long familiarity. Curt and I send each other Father's Day and Mother's Day greetings, and we share our love of Turner. Having the opportunity to love them all makes me cry with gratitude, with surprise.

The calm and fearlessness that I have had in the decades since working through that therapy feel like a return to my original strength, the strength from before the sexual abuse. The burning lesson remains, though: Even today, I reject the intolerable. There must be a solution and I will find it now. If I have a secret weapon or a tiny human-scale superpower, that is it. I was driven to conquer the depression and the anxiety, and now they have been gone for twenty-five years. If I could give a gift to anyone, it would be this rejection of the unacceptable, the unendurable.

IN THE SUMMER SEVEN YEARS AFTER I FINISHED THAT THERAPY, James, Turner, and I packed up ourselves and our dogs and drove the six hours out to Ithaca to visit my mom. My father had died the year before, and Mom was living on, determined and cranky at the age of eighty-two, in the house they had built together in the 1950s. Jim and Margaret, Jim's wife, had already been there for a week or more. We found Mom resting in bed, lying in the big California king my parents had finally purchased after decades in a double bed, eating chocolate-covered peanuts and watching TV. The big metal-framed window over her head looked out on the deep shiny green hillside of periwinkle ground cover rising up from the local-flagstone terrace she had laid almost sixty years before. She looked pale, and thin, and exhausted. This wasn't her usual depression and this wasn't a cold. She had a tremendous cough. She said her doctor had told her the pain in her ribs and other things were just side effects of coughing. But after a lifetime habit, she had given up smoking only a couple of years before. The next morning I took her to the hospital, and by the end of the day, we had the obvious diagnosis: stage four lung cancer.

The next year, Jim and I took turns flying from Chicago, for him, or from Boston, for me, to Ithaca, each of us for three or four days at a time, to take care of Mom. She needed someone to manage her many medications, and to shop and cook and clean, and she needed company. But she would not tolerate anyone coming in for her except the two of us, and the periodic visits of her hospice nurse. We cared for her this way for about five months.

Mom and I had our rough times along with the good during this period. No matter how I tried to tolerate everything, I could not swallow myself and my beliefs enough to let her bitter stories of all the people who had done her wrong slide off me. These were the same stories I had heard repeated my whole life. They were stories against my grandmother, and my aunts, and my father. Some of them were so terrible that I cannot write them. But many of them were just obviously ridiculous, like the anger that an aunt who had received an abundance of wedding presents had not passed any to my mother. Mom's bitter fury over this slight was as bright, brighter, in fact, that fall than it had been when this slight happened, sixty years previously. But in what world does a sister-in-law give away her wedding presents? The story made no sense. I do not know what terrible interior wound this story was a metaphor for.

Sometimes, I would ask her to stop. Then she would say, in a singsong tone of sardonic aggrievement, "Oh yes, Lindy doesn't like to hear any bad things. I must remember only to talk about happy sunny things with Lindy." I would feel a wave of rage and shame rise up and heat my face. Usually I was speechless. But finally, one day, some better logic kicked in for me, and I asked, "Mom, why do you tell these stories? They are the same stories you've told for years, so I wonder what is new for you. What can telling more do for you?" She was silent, looking down. I remember that we were on our way to help her change and bathe, and we were standing in the doorway of the bathroom off her bedroom. She was wearing an old pale blue chenille bathrobe and white terry slippers. She was standing in a moment of weakness, and somehow, also, a moment of honesty. She looked up and said, "I feel less alone if I can make other people feel the pain I feel."

IN DECEMBER I WAS IN SAN FRANCISCO AT THE AMERICAN GEOPHYSical Union fall meeting, leading a workshop about large volcanic events.

The room was full of scientists. All day people were reporting on our latest findings. Leonard Johnson, the visionary leader of the National Science Foundation program who funded us, was there leaning back with his arms folded, smiling and following along. A phone call came in on my cell. "I think the end is really coming, Lindy," said the nurse from the assisted care home Mom was finally in. "I think you should come right away if you want to see her before she dies." I flew to Ithaca that night, and I sat alone with Mom, just the two of us, for a week while she died. I gently massaged her feet and her hands, and brought her juice, and slept in the cot by her side, and finally, she faded and quieted and then she stopped breathing.

I called James and Turner, and Jim and Margaret. And then I went to the house of my friends Andrea and Amy. They cooked me my first pozole, the deeply warming and deliciously comforting Mexican hominy soup, and served me margaritas in their welcoming kitchen. They listened to me rant about my mother, and they kept me company as I got drunk. As I drank more I said more. All the stories just vomited out of me. I was the worst guest ever. I hope they have been able to forget what I said, and to forgive me.

I kept searching myself for sadness at her loss, for moments when I missed her. There were many such moments I had for Dad: I wish I could share with him the pleasure Turner and I have birding together, or a new plant I've identified, or a step in my career. I miss our summer barbecues and I miss learning what books he is reading. But in those exhausted months of aftermath, and in all the years since, I have not missed my mother. I believe that the depth of her blanket betrayal of me as a worthy human being, and the depth of her specific betrayals in not protecting me or believing me about my sexual assault, burned away any hope of love. And her own hurt was so deep that she had nothing left for me.

But I did experience a different, bizarre aftereffect: I suddenly felt freely myself, complete, and liberated. My confidence soared. I felt centered in myself. My work was productive, my voice was sure. And I felt liberated sexually for the first time in my life. My decisions were at last my own, and not hers.

Chapter 4

THE SEARCH FOR MEANING

What fictional concept or character do you most wish was real?" Using the primitive email system of 1996, I would occasionally send around little questions to my colleagues at St. Mary's College of Maryland. This beautiful tiny college stands alone on a green hill by the side of the broadening Potomac River, on its way into the Chesapeake Bay. The hundred or so faculty would lunch most days together on campus in an old house, now the faculty building. We'd talk and connect disciplines and knit together our small band. The day I sent that query, some people laughingly wrote back, "Flying reindeer!" and "The Silver Surfer!" But then our friend in the English department, Andrea Hammer, wrote, "Meaning." That took the laughs out of it. This answer rattled in my head from time to time over the years. Could life actually be devoid of meaning? Could meaning be a fictional construct?

After two years at St. Mary's I applied to return to MIT for a Ph.D. I had loved teaching at the college so much that I decided I needed that terminal degree so I could return to academia to stay. I sat to take the Graduate Record Exam right next to some of the senior undergraduates I had taught in math classes. I was a decade older than they were. One just looked sideways at me and said, "This is not fair." It didn't feel fair to me either; I was a decade behind. So it was with gratitude that I received an email from my master's advisor, Tim Grove, accepting my application for the Ph.D. program. I moved to Cambridge, Massachusetts, relatively near Lewiston, Maine, where Curtis had just moved. Turner went back and forth between Curtis and me. Sweet tiny Turner went to a year of nursery school and then to kindergarten in Maine with his father managing the infinity of challenges of being a single parent and a junior faculty member. Turner spent weekends and vacations with me, while I built myself back into a human being through the hard work of therapy.

I found an apartment in Somerville, a walkable distance from MIT. Making it a home for me and Turner felt urgent; though throughout our divorce Curtis and I communicated well and placed Turner at the center of care, it was, nonetheless, a divorce, and difficult for him. As for me, not being his primary caretaker during those two years still haunts me. I bought colored wooden letters and put his name on his new bedroom door. His bed was a nest of comfort: feather bed beneath, down comforter with a powder-blue cover above. His toys and books were there. When little blond five-year-old Turner came to the apartment for the first time, he saw his name, and he jumped into the fluffy bed, and we cooked dinner together. He was willing to give this arrangement a try, I felt. We walked our dog, Tess, around the neighborhood. My heart began to ease.

I tried to bring in all the best parts of my own childhood, and that definitely included animals. First, there was the stalwart and ever-present Tess, our red border collie who had kept our sheep in Maryland and competed in sheepdog trials with me, and who now adjusted with goodwill to life in Boston. She went to school with me at MIT as well as everywhere else we might go; she even met "Car Talk"'s Ray

Magliozzi when I went to pick up my 1979 Volvo from his shop after a repair. Ray took one look at Tess, came around the counter, and lay down full-length on the floor to talk with her and get licked all over his face.

Next was a short-lived hamster that bit us both. And then, one day at a good local pet store in Natick, Turner and I met an epic character in all our lives. After long minutes of deliberation, Turner chose from among several candidates a young black-and-white cowlicked guinea pig, and named her Domino. She lived with us for six years, dominant over the border collies, who would avert their eyes and slink away from her, and insistent that we bring new visitors to the house to meet her; she even earned her own theme song, composed by James.

At the time, in my determination to make a cozy, respectable home for Turner, I often spent my meager money on food or outings with him and then could afford only potatoes or oatmeal for myself for a few days at a time. We'd go to the Boston Children's Museum and buy ice cream. I'd buy fresh vegetables for dinner. And of course I'd drive the old Volvo up to Maine and back between Turner's dad's house and my own. And that would be all my money, joyfully well spent but finite. Eventually I pawned some jewelry to make ends meet; in my graduate program, we were not allowed to have a second job. I lived in a near-constant agony of feeling I was failing Turner, that I was on a precarious path to an uncertain career future, and that I was helplessly in love with James and yet not emotionally ready to be a good partner to him. So much uncertainty, so much worry.

One day, Turner and I were in my spare, shared graduate student office and Tim came by. He said hello, and Turner said . . . nothing. I could see his brain doing the meta-analysis: *I'm expected to act, there will be judging, there will be that common falseness of an adult making conversation with a child in front of his parent.* I could see Turner's discomfort, his anger, and his unwillingness to perform. I wished I could show Tim the happiness of Turner in his own element, playing with his stuffed animals, inventing worlds with his friends Annie and Lucy in their special clubhouse. Of course I felt the invisibility of his happiness was my own failing. I see now that it was both the blessing and the curse of Turner's intense awareness of human interactions and intentions. A blither child might have let more of the challenges pass him by, more of the questions and concerns go unanswered, but Turner needed to examine, experience, and reconcile them all. In a sense, he and I were going through the same process. And I was the only single parent among all the graduate students.

One day in the winter before James moved up to Boston, I said to Turner, "Would you like to come live mainly with James and me, and go to school in Massachusetts?" Turner became very still. Then he said, "How do you know that I can?" I'd already talked it over with his dad.

By then we'd been living in a little condominium in Acton for a year. We'd made it a home, sponge stenciling spirals and starfish around the downstairs bathroom and stringing holiday lights around the ceiling of Turner's bedroom. That spring of 1999, James and I were married in our living room. Turner and the other kids left partway through to go downstairs and play with the puppies Tess had had, and we heard the happy yips of puppies, the peremptory squeaks of Domino, and the conversation and laughter of the kids while we sat in a circle for our Quaker ceremony. Life was off to a fresh start.

SOON AFTER MY PH.D. GRADUATION, THOUGH, THAT BIT OF ASSUR-ance on the academic side began to erode. Early on in my time as a postdoctoral scholar at Brown University (a research position that typically lasts just two or three years, following a Ph.D.), I got a call from the University of Chicago, saying they wanted to consider me for a faculty position. I hadn't even applied! They flew me out and I spent the typical two days of back-to-back half-hour conversations with their faculty, and I gave two talks about my research, and I went to dinner with small groups of professors. I never heard from them again—not even an email. In this world of infinite distance and deep time and yet somehow without quite enough room for me, I began to wonder: What is a virtuous career?

I began applying for faculty positions. I had some constraints, because Turner was still young and Curtis and I wanted him to be near both of us. But Curtis was supportive of my efforts; he even offered to consider moving to where I found the right job. Sometimes I was short-listed for a position and given an interview, but I didn't get the offer in the end, with one exception: a lovely university that I had to turn down because of the challenges of moving family. I didn't know that would be my last job offer for three years.

They say, and I think it's right, that when you finish a high-quality Ph.D. you are the world's expert on your narrow subject. How can it be, then, that at the moment in my life when I knew the most about my topic and had become a clear, decisive thinker, I could not get an academic job? Could I actually have become less employable, when I had more skills? What did it mean to know so much about how our planet had formed, and then to be unable to find a place on it? And why did I want to be a professor, anyway? Our parents had bred into my brothers and me the idea that it was a deeply respectable and also pleasant vocation, with freedom, and travel, and gravitas. I began to worry that that was only really true in novels. Where was the meaning?

WORKING WITH MY COLLEAGUES AT BROWN WAS A PLEASURE, AND I wanted to stay. My main collaborators and advisors, geophysics professor Marc Parmentier and geology professor Paul Hess, created a warm culture enriched with an unremitting cascade of interesting research questions. Finally, that department had a professorship to hire for, and they were looking for someone in my field. I felt like the job almost had been written for me! I prepared my application with great excitement. An academic application is a bit of an endeavor. I updated my curriculum vitae with my latest conference presentations and a couple of new peer-reviewed research papers. I rewrote my teaching statement to explain how I could offer courses that fit with the needs of the department and would attract and retain students. I touched up my research statement with the latest grant I had won, and my ambitious plans to answer big science questions over the coming years, and then I submitted the whole package.

Applying for a job in the same academic department one is currently working in has many specific discomforts and awkward moments. The tenure-track and tenured faculty have written the job ad, and they have formed a search committee to assess the applicants and eventually choose a short list for interviews. These were the same people I was working with and sitting next to every day. Suddenly, there were uncomfortable silences around certain topics; suddenly, I had the twisty feeling that I was being assessed in new ways. One day the good news came: I was on the short list. That meant I got to interview with everyone in the department, and give one or two big lectures, and meet with students, and generally be assessed in every possible way over one or two full days.

I was waiting, full of hope, and even some expectation. I knew that at least some of the Brown faculty thought I was doing good research and that I was a good fit. I felt like I had an inside track. And then one morning, sitting in my office, I received a strange email. It came from a colleague at Brown. I opened the email, which had a subject line that referenced the job search, and I read with astonishment and an escalating sick dizziness a thorough critique and rejection of me and my work. A member of the search committee had accidentally sent this email intended for only the committee members out to the whole faculty and research faculty, a list that included me. That email enumerated in great detail why certain people should be considered as top candidates and why I and another person should be dismissed from the search. That is how I learned I would not be getting that job. I went back to working on my code for a while, but I couldn't hold the fiction that I felt all right working in my office at Brown anymore, and so I packed up my bags and went home.

In that dark moment, I didn't have much interest in applying for more academic jobs, but I was invited by colleagues at MIT to spend the next semester there, teaching an introductory geology course. Giving me this course to teach must have been an act of great sympathy on the part of my friends there, and I accepted with gratitude, fleeing Providence with some dignity intact. I loved teaching those twenty undergrads. Every day I remembered when I had taken that same course twenty years earlier, taught by the great John Southard. I had been enthralled, taking meticulous notes and trying to reproduce in my own notes his beautiful drawings of meandering rivers and sand dunes and the cross sections of volcanoes. Now, it was I who was drawing on that blackboard.

And then—in an outcome almost too wonderful to believe—the MIT department offered me a professorship. I had applied, not expecting anything, and was short-listed. I guarded myself, anticipating a replay of Brown. But here, though they had documented my every false move and imperfect outcome from freshman year on, they didn't care. Finally, I was an assistant professor. At MIT. My dream job.

WHILE AT BROWN MARC PARMENTIER ENCOURAGED ME TO FOCUS my work on the concept of the magma ocean. Early in the age of the solar system, the Moon, the Earth, and all the other rocky planets were melted by the energy of the giant impacting bodies of rock and metal that built them up in size. When a planet is a sphere of incandescent molten rock, it is said to be in a magma ocean stage. The magma ocean itself is the melted exterior of the planet, molten perhaps all the way through the rocky mantle of the Earth, underlain by a molten metal core. All our rocky planets were completely remade by magma oceans solidifying at least once, and in the case of Earth and Venus, probably several times, before the early solar system organized to the point that giant impacts no longer happened.

Our Earth's last giant impacts threw off material that re-formed into our moon in a molten state, and left the Earth similarly likely molten to its center. This moon-forming impact caused Moon and Earth to coexist with magma oceans. The Earth's magma ocean is far in the past: If the solar system's age is the equivalent of a twenty-four-hour day, that final giant impactor created the Moon and the heat of the impact engulfed the Earth in a magma ocean just thirty minutes into the day.

Humans have never witnessed a magma ocean, but the evidence from Moon rocks is indisputable. At some point, our Moon and our Earth were both spheres of incandescent molten rock, facing each other across an expanse of frozen space. At that time, we were ten times closer to the Moon than we are now. The heat of the molten Moon radiated across and struck the Earth, and the Earth heated the Moon in return. This coincidence of beauty, theory, and observation compelled me to work on this topic for years. The Earth's magma ocean solidified and as the solid minerals formed and less and less liquid magma remained, water and carbon dioxide bubbled out of solution and formed a dense hot atmosphere. The Moon's magma ocean was poorer in gases and also the Moon's small gravity cannot hold an atmosphere, so its tenuous gas envelope dissipated over a few million years, as the Moon began its slow retreat from the Earth to its current—and still increasing—sixty-three Earth radii away.

I'd been working intensively on the theory of magma oceans for about five years when, in 2007, I first stood in the open door of my newly furnished office at MIT, an assistant professor. I was a little stunned to find myself there. On the faculty. At that moment I was deeply appreciating the office itself, which would be my new home for many hours of many days. I had been given a small budget for furnishings. I bought a comfortable lime-green visitor's chair, and an extension of my modern, simple desk made a convenient table to work with students. The blond wood shelves were already filled with books, a state that makes me immediately feel at home. And I had my own giant window looking out over the river, just like the view I'd seen from Tim's office twenty-three years before.

The magical view of the Charles River from the tall Green Building at MIT had by then formed a kind of permanent imprint in my brain. I first saw it from the twelfth floor when I was a sophomore in 1984, beginning my very own research project in Tim Grove's lab. The institutional gray linoleum floors and bare I. M. Pei brutalist concrete walls were and still are matched by the gray steel desks given to each student. One whole wall of each river-facing office, however, is dominated by a giant window overlooking the glittering, sailboat-flecked Charles to the skyline of Boston. Every person's gaze goes out that window as they ponder their work.

Walking into my new MIT office each morning gave me a fierce blaze of joy. I'd settle into my desk and start working on computer models of magma oceans. While at Brown I had written a code that predicted how the magma ocean would solidify. This code, written on the MATLAB platform, consisted of about fifty routines that calculated the kinds of minerals that would solidify at each step, along with their specific and changing compositions; the density of those minerals and the composition and density of the remaining liquid magma ocean; the gases that the magma ocean liquids would release into the growing atmosphere of the planet; the temperature of every part of the atmosphere, magma ocean, and solid minerals; and the time each solidification step would take. So much study had been made here on Earth about how magmas solidify that I could extrapolate with some meaning even to magmas as deep as a planet.

The goal, of course, is to make testable predictions. With this code I could predict the resulting compositions and structures of the first planetary rocky mantles for the Earth, Mars, Mercury, and the Moon. Starting a couple of years before I moved back to MIT, and continuing for about a decade, I and my colleagues made great progress in predicting the initial compositions and structures of these rocky planets, and then matching the real observations from telescopes, meteorites, and primarily from space missions. By stepping these codes and predictions forward in sophistication and complexity, we've put meaningful constraints upon the processes that happened in that first blink of our planets' lives. We predicted from first principles the isotopic compositions of magmas on the surface of Mars, and the titanium content of tiny glass balls found mixed into the lunar soil, the result of fragmenting fire fountains of magma from lunar volcanoes three billion years ago. We were time traveling, surfing across those beautiful immensities of temperature, distance, and time.

This kind of computer modeling, though, has its limits. We had to make simplifying assumptions and try not to go beyond the most likely starting conditions and the best-supported physical and chemical processes. Generations of experimentalists, for example, have determined how certain compositions of magma solidify, and thus fed our models with the kinds and compositions of minerals. But after a magma solidifies past, say, 90 percent, that final 10 percent of remaining magma is such a strange composition that we can't reasonably predict how it will solidify based on existing experimental data. Thus, our conclusions had to be broad-brush, as well. We can't predict compositions or structures at small scales, or with very great certainty. After that decade or so, my research in magma oceans slowed, because the most useful and certain kinds of answers were either completed, or beyond our ability to predict. Back at MIT, I knew that moment was coming; I knew the magma ocean work was relatively low-hanging fruit, that it was a finite area of study, and that I had to be careful not to overstep what we could conclude usefully and with some assurance.

All this planetary-scale, long-timeline work on magma oceans was itself a reaction to work I had done in my master's and my Ph.D. Some of the work I had done then had felt, well, too small. In one study during my Ph.D., I had spent a year or so doing experiments on a peculiar lava composition that we collected from high in the Sierra Nevadas in central California. The result of months of sample preparation, and then a series of experiments, was the new knowledge of the composition, temperature, and pressure of the rock that had melted to make that lava. Amazing! We could actually labor away in the lab and learn what, deep inside the Earth, had partially melted to create the lava that erupted onto the tops of the Sierra Nevadas. That joy of scientific discovery was so great that it could disguise the fact that the payoff was too small for the effort.

To make my result about the Sierra Nevadas meaningful, I needed to connect it with all the other work that had been done on that subject, and then to connect the cumulative results with the Earth processes that made that melting and eruption happen. I had to compare and connect in a meaningful way all the other experimental studies on similar magmas, critique and either integrate or dismiss their results, understand how these mineral assemblages could have formed, and then, connect these compositional and chemical results and interpretations with an entirely different field, the physics of plate tectonics, mantle convection, melting, and eruption. Scientists are expected to connect their results with all the existing work on the topic, to compare, to put into context, and to synthesize conclusions from the whole. But stepping across disciplinary boundaries to synthesize from entirely different fields was even rarer a decade ago than now, and is always fraught with the dangers of misinterpretation and misunderstanding.

For the Sierra Nevada study, one obvious question was why did the melting source have no orthopyroxene or garnet, spinel, or plagioclase, the normal mantle minerals that melt to make magma, and instead have this mineral phlogopite, which is a surprise: it's a kind of mica. Mica contains water in its crystal structure. The other mantle minerals do not. Where did the water come from? And phlogopite, in particular, also has strangely high potassium and fluorine content. Where did those elements come from? None of these are common in the mantle—not the mineral, nor its water, potassium, or fluorine.

On the one hand, I was worried that the results I was getting were too incremental. Others had studied similar rocks from other parts of the world and gotten related results, some even including phlogopite. So in part, I was replicating work, but in another way, I was also adding, since mine was a new starting composition, and a more complete result over a broader temperature and pressure range. Was that enough, especially given the vast number of hours that every experiment requires, and the dozens of successful experiments that are needed to reach a conclusion about the origin of the rock?

We are always striving to take the biggest step we can that will still yield a confident result. Too big a step into the unknown, and you risk assuming too much, skipping intermediate questions, and missing information that would have changed your answer. You've now added doubt and nonuniqueness, if not actual incorrectness, into the stream of human knowledge. Too small a step, and your progress toward greater understanding becomes sadly incremental. I was worried that my experimental work was taking me to a state of incrementalism. And so, I added geophysics, the study of the movements of the Earth's interior and crust, to the results of the geochemical experiments we were carrying out. Instead of just running experiments in the lab, I learned in addition how to create numerical computer models of mantle convection and lithospheric dripping, so I could compare the results of experiments to the modeled predictions of where melting should be occurring.

The physics of the tectonic plates finally provided the answer. Other researchers had shown, using seismic waves to make tomographic images of the inside of the Earth, that the bottom of the tectonic plate under the Sierra Nevadas had sunk away into the deeper mantle. Some years later, at Brown, I made a simple computer model of the sinking lithosphere so that I could calculate its size and speed. Marc looked at one of my figures one day, and asked, "What happens to that rock chemically as it sinks?" That question was so clearly important that it felt like an alarm bell in my head, and I rushed back to my computer to calculate the temperature and pressure changes the sinking rock would experience. Knowing the stability of the minerals involved, we could then hypothesize that those sinking pieces had had their water, potassium, and fluorine squeezed out of them by increasing pressure as they sank. Those watery fluids reacted with mantle rock to consume orthopyroxene and spinel to make phlogopite; we actually had seen these reactions in our samples and experiments at MIT. That new, wet source melted, and the lavas erupted right onto the tops of the Sierra Nevadas, coating their white granite with a chocolate-brown layer of lavas.

For me, these studies were a turning point. Discovering the large-scale movements and processes of the planet, and eventually the solar system, was no incremental step. Previous researchers had suggested that volcanism would result when the bottom of the crust sinks into the mantle, but the specific processes they suggested turned out to be incorrect (the idea had been that the sinking piece left a dome shape in the bottom of the lithosphere, into which mantle would flow and melt by decompression, but fluid flow of even the lithosphere doesn't work that way, and the bottom of the lithosphere is left mainly flat after the drip falls away). By pursuing the physics of that dripping-and-melting process, I was later able to make predictions of the compositions and locations of resulting eruptions, and these were confirmed by other researchers. Together, we had identified and explained a new Earth process that leads to volcanic eruption.

How I loved this drip-melting idea, and the fact the phenomenon had been found in many places on Earth. I began to wonder, though, whether doing even this work was enough. The satisfaction of solving a problem and knowing more about how the Earth works was enough. But was the question itself big enough to merit the time and effort? What, as the sum of a human's contribution, is enough?

There is great beauty in the depth of knowledge humans have collected. I wish with all my heart that every person could, in at least one discipline, pursue and come to know through a long path traveled all that has been discovered, right to the edge of human understanding. Learning the knowledge landscape up to its outer limits bestows a perspective on what it means to be civilized, to know something in its entirety, to viscerally appreciate what it is to be an expert. This universe of knowledge is as complex, voluminous, and multidimensional as is our real universe, but the knowledge is less visible, is, in fact, largely invisible, until you search hard.

And yet, where should we each dedicate ourselves in the production of more knowledge? Was it worth it to understand how that Sierra Nevadan lava formed? If not, was it worth it to know how that lava was similar to others around the world, in the Altiplano, in Tibet, in East Africa? If not, was it worth it to know how all those lavas formed and were erupted because of chunks of tectonic plates sinking into the deep Earth? And if not that, then what was left to value?

In every generation, fields of study atrophy and vanish because progress is not interesting enough, or sponsors and funding sources change their priorities. Other fields thrive as new and bigger results are found. For me, moving to a larger physical and temporal scale of inquiry became an imperative. I had to be asking what felt like larger and more fundamental questions, though they are all still built upon the foundation of the chemistry of the planet.

WHEN HE WAS A LITTLE BOY IN ADELAIDE, AUSTRALIA, MY HUSBAND, James, was lonely. He felt his parents had made it clear in many ways that he was not central in their relationship, and he was an only child. There was more, and worse, that he waited day by day to leave behind when he was old enough to go.

At night James would lie in bed and think about the universe, and aliens, and communication between beings. I feel the resonance of this little boy yearning to create communication with distant, un-reachable, incomprehensible alien beings, who perhaps seemed none-theless closer and easier to reach than his own parents. To create communication, he reasoned, there had to be the ability for each side to recognize self and other, and the intentionality of each in trying to communicate. Self and other. One and zero, he reasoned. Binary. Any kind of intelligent alien would understand binary, he thought. And then, how to convince them that we are thinking creatures? Prime numbers. Send prime numbers, using binary. This about a decade before Carl Sagan wrote *Contact*, in 1985, and told the story of com-municating with aliens using prime numbers. For James, this idea was the fundament of how to reach, how to be acknowledged by, distant intelligent aliens. His immediate world was dire, but a seemingly infinite universe awaited, with creatures who would eventually speak his own language.

James now spends his days helping teachers all over the world bring joy and a human story into mathematics education. People ask him regularly if he always knew he was a mathematician. He was one, but he didn't know it for years.

Until he was about ten, he and his parents lived in an old house in Adelaide. The ceilings of this house were pressed tin, in the style of the late nineteenth and early twentieth centuries. The ceiling in James's bedroom was a five-by-five grid. He would lie in bed at night staring at the ceiling. As a little boy, perhaps age seven or eight, he'd count squares, and then count squares of two by two, and then three by three, and so on. And then rectangles. And then he began to trace paths.

If he started in the top left, he could march down the column and then up the next and down the next and so forth, and thus trace a path that walked through every square on the grid. True for every corner. True for the middle square. But seemingly not true for some other starting squares. Try as he might, there were starting squares from which he could make no path that walked through every other square. There was always a stranded square, left out of the path. Why? And was it really true? Perhaps he tried fifty different ways but always failed. Did that prove there was no path, or did it just mean he hadn't found the path yet?

Over many years, until he was in high school, James puzzled over this problem, while all the time he suffered through math in school. Math in school was all computations at speed. School math had no connection to the path problem he was still trying to solve; there was nothing in his experience to show that what he was doing in his spare time was actually mathematics. Math at school was all about *what*—what is the answer?—and the problem he was trying to solve was *why*. He had no one to talk with about it. His parents were not interested in his school life, or, seemingly, any part of his life. He quickly learned not to ask abstract questions of his teachers.

And then, one day while walking to high school, the visual proof popped into his head. He understood that the starting points of complete paths were like the white squares on a checkerboard, and the starting points of the incomplete paths were the black squares. He could see in his head, with sudden and beautiful clarity, why the paths from the black squares could not be completed. He had solved, for himself, a basic problem in parity, but it would be another five years before he would discover what math really was: not computation, but a field of beautiful whys, an endless landscape of exploration to follow, a theoretical landscape of clean, clear communications that, once they were discovered, were permanent and universal.

Though James was the first person in his immediate family to graduate high school, he knew even then that academics was going to be his ticket to a better life, and his ticket to escape the world he was living in. Like most Australians, he went to his local university, the University of Adelaide, and he lived at home with his parents. On Tuesdays, he would wake a little earlier than usual, and review and then memorize all the notes he had taken in lectures the day before. On Wednesdays, a little earlier still. By Friday, he had committed to memory everything he had been taught that week. His determination to find his exit ticket was unwavering, as solid and cold as iron.

And then, most of the way through a degree in theoretical physics, James discovered that abstract math was what his brain had been doing from the very beginning. His language of finding a way to communicate with someone, somewhere, who connected with him, his language of escaping his lonely bedroom, his language of escaping to a new life, was the language of mathematics, and it took him to the other side of the world.

GREGG AND I WERE WAITING FOR OUR LUNCH ORDERS TO ARRIVE, sitting in the bright window of the Trace restaurant in the W hotel on Third Street in San Francisco. We made a point of meeting for a meal every year during the American Geophysical Union meetings. We'd catch up on news from our families, and from work, and share news and predictions for the world of space exploration. I looked over at Gregg's calm, alert face with special concern, because he was in a long recovery from a terrible accident. I couldn't see the stress of it on him, though. His blond hair was as blond as ever, his skin tanned and healthy, his blue eyes steady and friendly, no wavering or exhaustion visible.

Gregg Vane and I met at the Jet Propulsion Laboratory (JPL) in Pasadena, where he was chief strategist for solar system exploration. I was working with JPL on the Psyche mission proposal, still in the early stages, just trying to get JPL to commit to partner and back the proposal. In Pasadena, and in Washington, D.C., Gregg and I had long conversations over sushi about the philosophy and the fact of space exploration. We reasoned through many thorny problems together: how to influence Congress to fund NASA at a higher level; how to position missions for greatest success; how to convince leaders of my organization and his to take key actions; how to plan years and decades ahead in space exploration.

Over the years I marveled at how even-tempered Gregg was. He'd quietly take in the details and complexities of the problem at hand, no matter how frustrating and complex, with his eyes clear and wide and his face relaxed. He was quick to laugh and never raised his voice or made emotional waves of any kind. And he moved forward with determination and clarity and had influence on all the problems he approached. This year, knowing the challenges he'd been facing, I hesitantly asked him if he'd be willing to talk about it.

"Gregg, I have a kind of a personal question for you. I wonder if you'd be okay to talk about it? You have been working seemingly tirelessly to overcome the devastation of that car accident. You've worked through months and months of rehabilitation. And your affect is as calm, happy, centered as ever. I am wondering—I've been thinking about how some people are resilient, and others not—are you aware, consciously, of the peacefulness you seem to carry with you? Do you know how you came to have it? I'm so interested in what allows some people to persist when others cannot. I'm fascinated. Is it too personal to ask?"

Gregg smiled. He said, "No, no, I do think about that! And I'm happy to talk to you about my family, and about how I got interested in astronomy."

With his same calm, personal, steady delivery, Gregg told me something about his childhood. When Gregg was young, his parents, with himself and younger siblings along, had moved out of town and into a stunning log house, built long before from trees on that land. His parents restored the house into a gleaming beauty. This image, the clean, bright house with its every knotty pine log individually sanded and polished, the project his two parents had done together, seemed to shine as the final positive moment of his childhood.

"Driving back from town one evening, we could see an orange glow from over the hill. It was our house, burning to the ground. Someone had taken a gasoline can from the garage, spread gas through the house, and set it on fire. It was arson," said Gregg, in his measured voice. And from that moment his parents' relationship began to unravel.

His father finally moved out, and his mother disintegrated into a state of alcoholism and began to dole out deep emotional abuse from the depths of a Jekyll and Hyde mental disorder that Gregg now thinks may well have been schizophrenia. Years before, when Gregg was six, he saw an exploding bolide one night (these are meteors that fall to Earth and make a fireball, but the shock pressure is too much for the material and it explodes) while driving with his father. He was instantly captivated and with his father's help began reading about astronomy in the library. Now, that love of the vastness of time and distance beyond the Earth came as a solace: on the way back from school each day, dreading what he would see and experience at home, Gregg would think about the universe and how tiny we are in it. Our tininess consoled him.

Why, I wondered, did that tininess of us, the seeming absolute proof of our meaninglessness within the universe, act as a consolation and not a desolation? Gregg simply answered, "I was always leaning toward optimism."

Gregg eventually went to college to study astronomy. He left behind three younger siblings, the smallest not yet ten years old. In college, Gregg saw people living differently, and suddenly, he recognized the dire circumstances in his mother's household. "You remember that parable of the frog in the pot? As the water heats up, there is no moment that alarms the frog and makes it jump out. Every moment is just an increment worse than the one before. That's how we were. But now, I was out, and when I looked back I realized how bad it was."

Gregg called his father, whom he'd had little contact with since he'd left the family, and explained that the lives of the other children were in danger. His mother drove drunk all the time, and could not be trusted to be safe with the stove, for example; and she had attempted to kill herself several times. There could be an accident, or another fire. His father had to get them out. Gregg's father said he would, but he wouldn't be able to do it without Gregg's help. And so Gregg had to go to court to testify against his mother, in front of the family and the judge. Gregg's father was made legal guardian, and the kids were saved from the immediate danger. Years later, he was able to reconnect with his mother, and was there holding her hand when she died.

I asked Gregg, "How do you explain your ability to transcend that childhood, to have a loving family of your own, to act professionally at the highest levels, and to be cheerful and present and steady? I don't understand how some people are crushed by experiences like yours, and others survive, and are able to love others, and function in the world." Astronomy, he answered. Looking out and beyond the Earth into something so great and vast in time and distance. Understanding our tiny part in the immensity of the universe. Seeing there is more than the emotions we carry in the moment of life we are in. "As soon as I learned about that," he said, "I felt strong and optimistic and motivated to be a part of that exploration. There's so much more out there than just us."

As with James's circumstances, Gregg's childhood could have crippled him for life. In both cases it would have been easy to explain away any degree of dysfunction on their parts because of their painful childhoods and lack of emotional support from their parents. Why do some children transcend their childhoods and become functional, motivated adults, and others do not? For James and Gregg, and for myself, it was the realization that we are only a tiny part of a vast unexplored universe. That perspective was both endlessly motivating and endlessly comforting.

Every time in my life as a scientist I have felt an irresistible thrill and an eye-widening need to pursue an idea or a project, it was because that project was a big new idea that stretched what is now into what can be in the future. Why do I love the idea of sending a space robot to a distant frozen asteroid, knowing it was once a hot young planetesimal, a tiny early-formed body, in the race to grow into a planet? For me the answer lies in how the geological sciences and the vast expanse of geological time and planetary development made the fragility and failings of human existence appear less dangerous and, in the end, less important. For James, the abstraction of math gave relief and perspective to the truth of his everyday life. For Gregg, the distances between galaxies soothed the pain of the tiny moments of the everyday. For me, time is the great soother: over billions of years, none of our mistakes mean anything. For each of us, these experiences were as authentic as it's possible to be. They were visceral, defining, and they built bridges to our futures, and between ourselves and others.

And for me, finding the solace through education that allowed me to move forward in life was the necessary first step to asking, Why am I seeking solace? Taking those steps, asking those questions, was necessary to prepare for greater action and greater change and greater meaning later. I may have gone into natural science to steep myself in a comfort and to soothe a fear, but the fascination with the natural world was real and lasting. I began to study geology in college in part as a reaction to hearing Hans Bethe talk about the end of the human race. With each step toward understanding and mental solace, I was building a foundation on which I would later erect new knowledge, new jobs, new relationships, and eventually, a NASA space mission beyond Mars.

Inevitably, though, the question of value and meaning pressed me, and my family, beyond science. If I thought there were parts of science with deeper meaning than others, then surely there were whole fields of endeavor more steeped in meaning than others. Turner, then a young teenager, and James joined me in thinking through this question. We'd query people on social media or chat with friends and neighbors from time to time, always asking, "What is the most meaningful thing you can do with your life?" We began to explore the notion of a virtuous career, making meaning of the way we spend our time. In the beginning, we thought of some of the stereotypical answers: The ideal could be Florence Nightingale, someone who works every day selflessly to alleviate the suffering of individuals. Perhaps scaling that concept has more meaning: find a way to raise money for food, or clothes, or medical care. Or perhaps the view that true suffering is elsewhere is condescending and Western. *What about fixing our own body, doctor? Making shoes for our own children, cobbler?* These, however, did not seem to be the right answers. They felt, at the time, like Band-Aids, not like solutions, not something lasting.

One conversation went on longer and went deeper, and we focused on the idea of scalable solutions. As usual, we were hanging around in our living room in the apartment that St. Mark's School, outside of Boston, where James was teaching, had provided us. I was sitting on the orange couch with my laptop, with Turner across from me on the other couch, and James was working at the dining table near the terrarium Turner had built for a pair of brightly colored dart frogs.

"What is it about the charity model that seems wrong?" I asked.

Turner immediately replied, "Whatever problem needs solving has to be led by the people who are experiencing the problem."

"And the solution needs to be systemic and lasting, not just a Band-Aid for once," added James.

Ideally, we all agreed, the solution needed to extend beyond one instance of the problem to many instances around the world: it had to be scalable. We were taking the issue that had troubled me in science and were applying it to the real world: find solutions that solve more problems at once.

We decided, that day, that the answer had to be education. We had to make education relevant to the age we were in, so that people were able, motivated, and filled with the belief that they could solve the problems that faced them, whether in their family, community, or society.

We were pretty pleased with ourselves! We had our answer. This was around 2009, and Turner was a senior in high school, and James was already forging his own path toward re-forming math education to contain joy and a human story. There our biggest answer to meaning—education—sat for three or four more years.

Chapter 5

EVERY ENDEAVOR IS A
HUMAN ENDEAVOR

Most volcanic activity on Earth occurs along the boundaries between the tectonic plates. Either one plate slides down beneath its neighbor—a subduction zone, as in the west coast of South America, and the Philippines, and Japan—or the plates move apart and create new crust, as at mid-ocean ridges. But periodically, just a couple of dozen times in Earth history, a vastly more voluminous eruption has occurred: a flood basalt. These flood basalts make the little eruptions I had studied in the Sierra Nevadas look like eyedroppers. During flood basalt episodes, fissures open in the crust and lava pours out, year after year, for about a million years, until an area the size of a country, or a continent, is covered in black basalt, which is the name for the rock that results from cooling that low-silica lava. The most recent flood basalts in Earth history, the Columbia River flood basalts, erupted the majority of their volume about sixteen million years ago, and covered much of Washington and Oregon and parts of Idaho. But their measly 42,000 cubic miles of lava is only about one-twentieth the volume of the Siberian flood basalts that I'd spend several decades attempting to understand.

In the first year of my Ph.D. at MIT, I had taken on a research project about flood basalts, working with Brad Hager, a geophysicist who eventually became a co-advisor of my Ph.D. Specifically, I was studying a flood basalt that occurred 252 million years ago in present-day Siberia. This flood basalt was perhaps the most voluminous to ever erupt onto a continent (though there are larger ones that formed in and upon ocean floors). The Siberian eruptions produced more than a million cubic kilometers of lava, perhaps as much as four million cubic kilometers. If they had erupted in Kansas they could have covered the entire lower forty-eight states in lava five hundred yards deep.

The Siberian flood basalts seemed, with the best age dating science of the day, to have erupted at the same time Earth experienced the worst extinction in its geological history: the end-Permian extinction. In the end-Permian extinction more than 70 percent of terrestrial species and more than 90 percent of ocean species went extinct. And no one knew whether the flood basalt and the extinction were related. This volcanic event was perhaps the largest to occur on land in the history of the Earth, and this extinction was definitely the largest in Earth's history. Causation or coincidence? The question boggled me: How could we not know? And then I discovered we didn't even really know why flood basalts happened. All questions, no answers.

Perhaps it seems self-evident that a flood basalt would cause extinction. "Volcanic eruption" tends to call to mind volcanoes in Indonesia or the Philippines, or Mount Saint Helens in Washington state: Explosive eruptions create ash clouds rising high into the atmosphere, and produce superheated gas clouds that rush down the volcano's flanks. The towering hot plumes of gas carry climate-changing gases into the Earth's stratosphere. Flood basalt eruptions, however, are hardly ever explosive. Without the explosive plume, gases seldom get transmitted worldwide.

The lava of flood basalts oozes out much as it has been doing in recent years from Kīlauea in Hawaii. Unless you are unlucky and trip or end up breathing too many fumes, you can go right up and look at it and go home alive. Many scientists thought that such an eruption could not drive a global extinction. No obvious mechanism could be found for such an eruption to cause animals and plants on the other side of the Earth, or in the oceans, to die. The most effective way, perhaps the only way, to cause extinctions around the whole Earth is to change the chemistry of the atmosphere. Since flood basalts were thought to be very seldom explosive, people thought they couldn't plausibly cause global change.

Brad and I started by thinking mainly about how the flood basalt itself formed. How did such a giant volume of molten rock appear under a thick, cold continental plate like Siberia's?

Magma (which is called lava when it erupts) results from melting parts of the Earth mantle, the vast volume of rock between the Earth's crust and metallic core. On the surface in our human world, if I asked you to melt something, you would probably heat it up. Here on the surface, changing temperature is the way to melt something since we exist at a more or less constant atmospheric pressure. But inside the Earth, temperature changes are so small and so gradual that they almost never cause melting. Inside the Earth, little temperature changes cause the solid rock to convect, that is, to move, like boiling oatmeal in slow motion.

The hottest rock is just above the metal core of the Earth. Heat causes materials to expand a little bit, and that bit of expansion, about three parts in 100,000 of expansion for each degree Celsius a rock is heated, makes that rock a little bit less dense than its cooler neighbor. The hotter rock will rise up because of its lower density (as a helium balloon rises in our nitrogen-dominated air), while the cooler, denser rock sinks down. Even though all that rock in Earth's mantle is solid, it will flow like a liquid on geologic timescales. The hot, pressurized mineral crystals will deform and move past each other slowly, more slowly than a snail, about as slowly, in fact, as your fingernails grow. The hot rock moves upward, and rock that was near the surface and cooled by conducting its heat away into our atmosphere and out into space will just as slowly sink back downward.

So inside the Earth, temperature change is seldom the way rock melts. The way the mantle melts is usually by moving upward and thus being depressurized. Inside the Earth, each parcel of rock experiences the pressure (the weight) of all the rock between itself and the Earth's surface. If a parcel of mantle flows upward, there is less rock between itself and the surface, and it experiences less pressure. Removing the pressure from a parcel of hot rock can allow its volume to change from solid to liquid, that is, to melt.

The mantle can only flow upward, and thus reduce its pressure, until it reaches the bottom of the crust, or, to use the correct term, the lithosphere of the Earth, which is cold and stiff and doesn't flow like the mantle does. The lithosphere of the Siberian continent is, and was even 252 million years ago, very thick—around 60 to 120 miles—and thus extends deeply into the mantle like the keel of a ship. Any mantle rising under this thick lithosphere couldn't rise as far as it would under, for example, thin oceanic crust. How, then, was the mantle able to depressurize enough to melt so very much, and erupt so much lava?

This question led to my first interesting scientific hypothesis, some years before the drip magmatism idea.

That hypothesis was that the mantle under Siberia melted only a little bit before its upward movement was stopped by that thick Siberian lithosphere, and that melt then percolated up into the lithosphere and froze there. What would that freezing magma do to the lithosphere? The act of freezing releases heat into the lithosphere, and a warmer material flows more easily than does cold material. The frozen magma is also denser than the lithosphere, so its addition makes the lithosphere as a whole more dense. That bit of changed lithosphere would then drip off the bottom of the continent and sink into the mantle. With enough of these drips, the lithosphere would be thinner, and the continuing upwelling mantle able to depressurize and therefore melt more and more, and to produce a flood basalt.

As with all scientific ideas, they rise out of the collective consciousness: people had thought of drips on the bottom of the lithosphere before, but I added the physics and chemistry of the freezing magma, making the lithosphere warm and dense and therefore to sink, which would allow more melting, and the connection to the flood basalt.

That paper became a part of my Ph.D. thesis, and after my Ph.D. as postdoctoral fellow at Brown University I kept working on different parts of the Siberian flood basalts and end-Permian extinction problems. One day my great friend and mentor Sam Bowring, a professor at MIT, said to me, "You've been thinking more about the Siberian flood basalts and the end-Permian than almost anyone else recently. Why don't you put together a team to really figure that out?"

As those kinds of calls to action sometimes do, Sam's question rang loud in my head for some months until it began to feel real. Finally, with Sam's encouragement, I summoned the courage to telephone Leonard Johnson, a program officer at the National Science Foundation. I asked Leonard whether his program might fund a workshop on how to solve this big important problem. He said he would! And so I, a mere postdoc, set about inviting the luminaries from this field.

I felt I was punching above my weight, calling the National Science Foundation out of the blue, inviting senior researchers to drop what they were doing and come to a workshop; even setting up and running a workshop felt out of my league. We ended up with about twenty-five attendees from several countries, including France, England, Norway, and of course, Russia. Western scientists did not have enough collaborations with their Russian counterparts, and these critical collaborations are hard to establish if you do not meet the others in person.

On the first morning of the workshop at Brown, I stood up in front of my guests, every single one in permanent jobs while I was still a junior impermanent postdoc, and I asserted that to make progress on how the Siberian flood basalts formed, what caused the end-Permian extinction, and how they were related, we needed to put together an interdisciplinary team that pursued questions no one discipline could do alone, and that shared and discussed data across those boundaries. The people studying the single-celled organisms that lived in the end-Permian oceans did not spend a lot of time cross-correlating their data with the physicists who made models of volcanic eruptions. Neither of them talked much to the chemists studying the Siberian rocks.

We also needed samples of the rocks themselves. By analyzing bubbles and minerals in the rocks, we could discover if the lavas had carried climate-changing gases after all. By analyzing elements in the minerals, we could also finally pinpoint the time of the eruptions, and then know if they truly coincided perfectly with the extinction, or whether the extinction started before the eruptions and thus had to be caused by something else.

The collaboration we were attempting didn't appeal to everyone. Some people didn't end up joining our project. And one person came to that early workshop, listened to all our ideas about what we could do to solve the big Siberian mystery, left the workshop, never replied to another one of my emails, and then did two of our proposed projects himself. But you know the best part? It didn't matter. Because the other 80 percent of us made a great team and we went on to do so much more.

I worked with the team who stayed committed after the workshop and we submitted a big proposal to the NSF to work on this project. The proposal was more than twice the length of a normal science proposal, and at the time I felt the weight of its thirty-six pages, two dozen collaborators, and lengthy budgets and financial descriptions—though this burden would dim significantly when just the first proposal for the NASA Psyche mission reached over two hundred pages.

We waited the requisite months for the NSF to go through its review process. We didn't get the funding. Resubmit next year, we were told. The team was generally upbeat; funding rates are usually between about 10 and 30 percent for any given year, and this program almost never funded a proposal the first time through. I felt good, actually, knowing we had put together a credible proposal. We collectively shrugged our shoulders and turned to other work in the meantime.

During that hiatus, I was hired as an assistant professor at MIT, and so the next round of proposal work commenced there. I'd had a year to think about teams and how they worked. I began to pull from the experience I had had at the helicopter factory almost twenty years previously, when I was fresh in the workforce after my undergraduate degree. How could we let ego and reputation go long enough to really collaborate? In other grants, I'd seen each senior member of the team take their funding back to their own lab and use it to tackle the obvious next step on the problem within their own field. I think of this as taking your slice of the pie home and doing what you would have done anyway.

Each person in science has their own subspecialty, a narrow sliver of subject matter about which they might know the most of anyone in all the world of science. Because careers in science are often made by having some good ideas that other scientists begin to follow and then build upon, strength of character can play a significant role in creating a career. I knew I could have a great idea and write a great paper and have it reviewed by peers and published in a scientific journal, but if I was not out there at conferences and other universities talking with colleagues and arguing through the fine points and comparing my idea to theirs, I was unlikely to build up an intellectual leadership in my field. I'd been advised that if I didn't make some other scientists angry, I wasn't clearing a space for myself in the field. I used to joke that in some small subspecialties, researchers were known for circling the wagons and shooting inward; that is, they wouldn't even support competition from their closest collaborators or in some cases even their own graduate students.

And when people in the same field of study have known each other for decades, jealously watched each other's successes and also silently cheered their failures, the disagreements can be all the more personal.

While we were writing the proposal, I had two people in closely related subdisciplines on the team. When one of them reviewed the proposal, everywhere his competitor's name appeared, he crossed it out and wrote his own. When I got the document back with track changes showing all those crossed-out names, I could not believe my eyes: Were we still in kindergarten? And there was more: big proposed budget increases, and at the same time reduced work responsibilities. I could only imagine how poorly communication would go after we were funded, if this is what happened before we even had a project to work on, and knew I had to remove him from the team.

The problem was, he was a famous senior scientist, and I had not yet earned tenure. If I made an enemy of him, he could really damage my career. I called Leonard Johnson, the NSF program director who would be receiving this proposal. I told him what I was about to do and why. He received this information calmly. We didn't know each other well. I hoped I had done the right thing. I didn't want to bring unnecessary drama, and also didn't want the program director to think I was weak and needed protecting. I was attempting to be strategic.

My hands were trembling and I was sweating when I called to tell the team member he was off the team. The sound of the phone ringing in my ear seemed unnaturally loud and slow. The call was brief. Later, I heard that he had indeed immediately called the program director with the complaint that I was unsuited to lead, and to my very good luck, the program director respected what I had done. But that was the end of it, and after a few years of cooling off, this former team member and I can even enjoy chatting when we meet at conferences. If only every team fix turned out this well.

Every research endeavor, it turns out, is a human endeavor. Every endeavor involves people interacting, competing, and, we hope, collaborating. Every endeavor involves people reacting and forming their own conclusions. The stereotype of a lone man in a lone lab making brilliant progress is a lie: every scientist needs to work with, contend with, and convince their community, and also, it's so often now a woman and not a man. Connecting with and convincing others is a big part of driving knowledge forward, and along with this come the attendant risks of ego, distrust, secrecy.

I was attempting to connect the people through the questions we were asking such that they had to work together and create a greater common result. I drew figures for the proposal showing how the results all worked together to build greater conclusions. I budgeted for graduate students and postdocs who would be cross-appointed among two or more lab groups. It was a start.

As we were beginning to write this second version of our proposal, I was introduced to someone going into the field in Siberia. I had some money left over from the initial workshop the NSF had funded and

Leonard Johnson gave me permission to use it to make a reconnais-
sance trip.

I was ecstatic. I took an intensive Russian language course; my
studies ended just as we were learning a kind of future tense verb that,
my teacher explained, was used for things you didn't really think were
ever going to happen. I bought a new backpack, and worked out, and
then when the time came I flew by myself to Moscow.

WE RUSHED ALONG MOSCOW'S FIVE-LANE-WIDE RING ROAD IN THE
corporate car of Norilsk Nickel, the Russian mining and smelting company: Valeri Fedorenko, who was the world's expert on the Siberian
flood basalts, the driver, and myself. I was sitting in front. I reached up
with my right hand to pull the seat belt across myself, but I felt a gentle
touch on my hand stopping me. Valeri quietly spoke in my ear that a
seat belt was an insult to the driver, and we knew they only trapped
you and did not make you safer, anyway. So while in Moscow, I rode
without a seat belt. Valeri had learned his English mainly from reading detective novels, he told me. The driver gestured proudly toward
a copse of trees at one point, and told me, "русские березы," *Russian
birches,* the beautiful white-barked soul of the Russian countryside.
Later, Valeri and his driver stopped for a cigarette, and explained that
smoking was excellent for your health. What does it mean to "know"
something? I asked myself.

They dropped me at the gatehouse of what looked like a disreputa-
ble apartment block arranged around a central lawn. The guardhouse
interrupted a high, spiked fence that surrounded the apartment block,
which I learned later was a hotel and conference center for government employees. This was not one of the new glamorous hotels that
began to rise in Moscow along with the oligarchs and the oil money;
this was 2006, and not in the center of the city. I was the only foreigner
I saw. The guardhouse was only large enough for walk-through and

tabletop metal detectors, a desk, and two guards with semiautomatics. They did not smile a welcome. I passed all my backpacks through the detector, and walked through myself. They checked my passport against a handwritten list, and after examining my rock hammer, Leatherman, and cold chisel, impersonally pointed me toward one of the high-rise buildings.

My rudimentary Russian was already coming in handy. Throughout the five field seasons I'd spend in Russia, I encountered few Russians beyond our field group and scientific colleagues who spoke English. I liked the feeling of the Russian words in my mouth: "на право" (*na pravo*) is what the guards told me. *On your right*. The thrill of these small comprehensions was mostly caused by the contrast of my incomprehension of most of what was said.

At the airport the next morning, Valeri kissed my hand good-bye and smilingly accepted a series of gifts I had brought for him. I was flying to Norilsk, the second-largest city inside the Arctic Circle. I fell asleep as the plane took off, and awoke a few hours later to see dawn over Siberia at 1:30 A.M. Moscow time. The flight attendants had found one of their number who spoke English, and she asked me if I wanted chicken, beef, or fish, and when I would like dinner. They brought me an elegant slim glass of apple juice and covered my tray table with starched linen. Dinner was lovely: a hot chicken cutlet with rice; a plate of smoked whitefish and salmon; some cold meats; a dish of salad with some of that incomparable crunchy, flavorful Russian cucumber; fresh grapefruit, kiwi, and orange; a roll; a slice of Russian rye bread; a petite lemon tart; an elaborately wrapped chocolate; truly delicious coffee; and later, a fresh pear from a basket. The dinner brought tears to my eyes, and I thought, I must be very lonely that this kindness makes me cry.

The plane was approaching Norilsk. One of the crew members came to me and said, politely but firmly, "сидеть" (*Sit!*). I understood correctly that he meant I should not move until someone came for me. We landed, and everyone else in the plane filed off. I sat. When the plane was empty, the man came back and motioned for me to follow him. We walked down the steps off the plane and got into a kind of open-topped Russian jeep. Chatting cheerfully in simple English, he drove me away from the terminal to a small wooden shed. He took my passport and left me sitting by myself in a room with a bare desk, two chairs, and a poster of Russian military uniforms. I waited. I wondered when I should get nervous.

The man came back and sat at the desk silently filling in a form. Then, someone began knocking on the outside of the wall, calling, "Американка? Американка?" (*American? American?*). My soon-to-be friend Alexander Polozov leaned down outside the shed and fit his face sideways into a little window the man made by sliding a board back. "Leenda, Leenda?" he asked, and I said, "Yes!"

In the airport terminal at last, I met Sverre Planke, a professor at the University of Oslo and expert in flood basalt interactions with sedimentary rock, who had invited me to join this trip. I could not believe he would welcome a stranger into his group; this was a huge stroke of luck for me. I later asked Sverre why he was willing to let me join them, and he said, "Not just anyone is willing to travel to Norilsk by themselves, and besides, if you had been a pain we just would have sent you home." The kindness of Valeri and of Sverre and Henrik Svensen, also from Oslo, changed my life. Picking the team, in science as in dodgeball, is everything.

The next morning, we rode through town in taxis to get to the headquarters of Norilsk Nickel. The main street unfurled clean and impressive, with imposing multistory official buildings. One block away, buildings were deteriorated, and along the boundary roads stood the brutal rectangular concrete apartment towers I was to see all over Russia. Some of these were half built, tilted over, abandoned when their foundations failed in the permafrost. A dark metal frieze covered the side of a building at a corner: a memorial for the prisoners who died in the Norilsk Gulag outpost. These prisoners had built the town. The person who ran this Gulag attempted to save the lives of some prisoners, we were told, by having anyone with any geologic expertise spend their time prospecting rather than in hard labor. In this way, some of the improbably vast mineral wealth of Siberia was discovered and cataloged by ragged men who traveled with reindeer and chipped through the moss and shattered permafrost with pickaxes to discover what rocks lay beneath.

When those almost unimaginably voluminous magmas intruded the Siberian crust and began reacting with the existing rocks, some of the magmas mixed with an excessive amount of sulfur and produced a special suite of minerals rich in precious metals. Thus the alchemy of heat, volume, magma, and crustal rocks produced both an environmental disaster unmatched in Earth's history, and a mining boon of equal proportion. In the office of the chief of drill core storage for Norilsk Nickel we had cognac and chocolates. He told us that the motto of the Siberian geologist is "No snivel." In the distance, we could see the tall buildings that housed the mine heads, where ore continues to be brought up, year after year, and smelted with convenient local coal.

That night we were invited to the house of Viktor Radko, the Siberian geologist for Norilsk Nickel. Viktor lived in an apartment in one of the concrete blocks. We strolled through town in the blinding horizontal searchlight of the Arctic nighttime sun. We walked up the utilitarian bare staircase, metal and scabrous concrete steps, smelling of stale humanity, and passed through Viktor's double steel front door with its metal crowbar that slotted into a custom socket in the floor, holding the door shut. Alexander told me that this was normal. What could the endless Arctic winter night be like here? I wondered.

Viktor was tall, thin, and strong, an archetypal field geologist, with a warm welcoming smile. I offered him my limited but heartfelt Russian greeting and he accepted it kindly, and had similarly little English to offer in return. Over a lovely supper of potatoes, chicken, cucumber and cabbage salad, smoked fish, bread, and wine, we talked about geology. Viktor spent several months alone every summer doing fieldwork throughout this region of Siberia.

What was the most dangerous thing in the field? I asked. Moose, far more dangerous than the big brown bears. And what did he bring with him? I asked. Why, his tent, his gun, his fishing rod (he had to feed himself, being unable to carry food for those months), and his cat. His cat? I asked, thinking I had misheard. Yes, Russian geologists sometimes bring their cats with them in the field, as Americans sometimes bring their dogs. He showed me a photo of himself in a vast tundra, holding a large gray cat in his arms.

Viktor gave me a gift, a tiny jewel box made of polished slices of Siberian lavas that had been altered over the ages by hot water flowing underground, into the dark green, patterned rock called serpentine. I carried that box home so carefully, through all the adventures that were still to follow. He then offered us some special Armenian cognac, and some black Russian bread. I was ready to pledge "на здоровье" (*to your health*), but Viktor wanted to teach us a genuine toast. "Hold your shot of cognac in one hand," Alexander translated, "and pick up a chunk of heavy fragrant black bread in the other. Breathe out, out, until all your breath is gone. Then take the shot of cognac! Quickly, hold the bread to your nose and mouth, breathing in only through it, tasting the cognac and smelling the bread, and think of Mother Russia."

For tens of miles around Norilsk the taiga is dead, killed by the exhalations of those smelters. The old larches stand as gray skeletons scattered across the rolling hillsides. That day I saw my first long Siberian vistas. A week later, I would see them for the first time from the porthole in an Ми-8 helicopter. This big troop-carrying helicopter flew so straight and steady that I could open the porthole and lean out from my seat-belt-free bench and see the Siberian taiga, flat, flat, flat, out into the dimmest blue distance, dotted with swamps and rivers and the hardy small trees of the Russian north, so vast I felt a little bit unnerved. This wilderness felt too big to fit into the Earth I knew.

These endless bogs, winding rivers, and scattered larch forests completely covered the lavas of the distant past. Only rivers cut down through the bog and rock to reveal the rocky landscape of the past. In 2006, in the weeks after that day, we traveled about six hundred miles in Russian helicopters, and saw only one rock outcrop that was not on a river. To see the rocks and sample them, we had to take to these rivers.

I wanted, I hoped, that eventually, as part of this project if it was funded, to find the rocks that resulted from the early explosive ash and gas eruptions that predated the lavas. Those early explosive eruptions could have sent the climate-changing gases high enough into the atmosphere to affect the whole globe. But the rocks were almost mythical; so far I had not found any person who had seen them, though they were drawn on one large-scale geologic map, in an area far to the south. We'd be going near there next and my hopes were high.

Once we were in much more southerly Bratsk, we shopped for food, and we awaited clearance for our helicopter to take us to the next geological site. I recorded new Russian vocabulary in my field notebook. *To wait. To camp. Rain. It is raining. Okay, we agree. Bag. Box. Larger. Stronger. Tape. To carry. I am sorry.*

A few things give me that same deep intense pleasure and thrill—a fast roller coaster, the roar of a rocket sending something into space—of leaving the Earth behind as you rise like air in a helicopter. Oh, the magic of helicopters. In an hour, after the vistas of winding rivers, forests, and bogs, we were setting down in a wildflower meadow under a low gray sky. A few abandoned wooden buildings stood around. Once our helicopter had departed, everything was silent.

The grass and wildflowers were so dense we were unable even to pitch our tents. Henrik made a clever tool: He held in his hands ropes that were tied to each end of a flat board, which he stood on. He pulled up on the ropes as he swung the board and his feet forward, first on one side and then the other, and in that way he pressed the grasses flat as he waddled forward. Our camp became a little clearing in the big wildflower meadow.

With numb hands in the cold drizzle we chopped down saplings to allow the old doors of the sheds to open, and to make space to carry the core boxes out. When a mining company drills rock cores, continuous cylindrical borings of solid rock from promising underground formations, the core is laid into endless numbers of wooden boxes. These boxes, a bit like kitchen drawers, hold 6 or 8 or 10 sections of core in tidy continuous lines, like writing on a page. The top of one core had previously been connected to the bottom of the one lying next to it, so on through that core box, and into the next, and the next, until the hundreds of meters of rock core extracted are complete, and then the next core begins. These boxes were piled in strong shelving in huge storage sheds, like the very library of the Earth. Over the years we were to see thousands of these boxes, the fruit of decades of labor, holding irreplaceable information about the rocks that underlie the moss and spruce through all of central Siberia. Sverre and Henrik had read the documents and knew exactly which rock formations these drills had cored, and which samples they needed for their work.

Years previously, an airplane flying over with a magnetometer had shown this area of the taiga to have a strong magnetic signature, meaning, probably, a big iron deposit under the thick trees and moss and bog. Later, a team of men came, cleared trees, used the trees to build housing and lab space and sheds to hold the many core boxes, and then they built those core boxes, and they drilled deep cores, and retrieved them, and analyzed them, and packed them into boxes and piled them to the rafters of the buildings, and when they were finished, they left it all behind. Now, years later, we came to an abandoned clearing like hundreds of others around Siberia. In 2006, the buildings had not yet all had their roofs collapsed by snow and had not been run over by forest fires, so the richness of the cores was still accessible. The area was scheduled to be mined starting in 2010, so all I describe here may very well be gone.

We laid the boxes over the flattened meadow grasses and flowers, which were sweating out the lovely smell of their juices, and we picked out the fascinating rocks and recorded them in our field notebooks, and wrapped them in plastic, and labeled them inside and out.

A Russian man with a bright bandanna on his head walked out of the woods and up to the men of the group, startling us, studiously ignoring me after a brief glance. It was an encounter I knew well from the field—that notion that as a woman I belonged to the men, and so he could not look at me without causing tension with them. In these moments I felt myself diminish into tininess, or into a symbol for sex rather than as a whole person. This man was out berry picking, he explained, and asked with interest what we were doing. After a brief conversation, he flashed a smile made mostly of gold teeth, and disappeared back into the forest. I do not know how far we were from a village, but often in Siberia that did not seem to matter.

The temperature began to drop, and it was time to make dinner. Alexander and Henrik started a fire and we set up our tents, and dragged some tree trunks and boards near the fire to sit upon. That finding-of-seats, a perpetual camping ritual. Alexander made us tea with wild currant leaves. We cooked something in a pot for dinner, and afterward, I offered to wash dishes. Alexander, in all of his kindness and deeply ingrained chivalric belief that I could not function on my own, went down to the river with me and showed me how to clean the bowl with fine sand, as if I had not camped many times before. Alexander, and the Russian men I later went into the field with, were impeccably polite and friendly but also seemingly incapable of letting me do, make, or carry anything other than food. With my American and European friends, this was never a problem; Sverre was happy to let me carry my own pack, haul core boxes, pitch my own tent. But to the Russians I was a different kind of person.

Snow began to fall. We sat around the fire in the snowy dark and sipped some vodka someone had brought, and we told stories and laughed. I felt myself again. But many other times in Russia I felt I could not be myself and at the same time a woman. I was so different from the miniskirted, stilettoed women of Moscow, and so different from the scarfed shopkeepers of the Siberian towns, that I often felt I was some third sex.

Tonight, at least, we were one team. In the coming months we'd rewrite the proposal and this time we would win. But now, we sat companionably around the fire, knowing we had our samples, and we'd be picked up by the helicopter the next day.

WE WERE OFF TO RUSSIA AGAIN IN 2008, FUNDED, OFFICIAL: SAM, the geochronologist who determined the dates the rocks formed; Brad, the geophysicist who worked on the fluid dynamics of the Earth and how the flood basalt could be produced from the mantle; and me and Ben, the igneous petrologists, learning the composition and formation processes of the magmas.

Earlier that summer, Ben Black had come to MIT to join the Siberia project and earn his doctorate. Officially, he was supposed to start in September, but he couldn't bear the idea of missing our first field season, in July 2008. I worried about bringing someone I barely knew to locations as remote as where we were planning to go, out on the

Kotuy River, above 70 degrees north latitude in central Siberia, hundreds of kilometers from any settlement. I made him promise never to whine. He swore he never would.

We proposed to do three critical scientific investigations. First, we would create a high-precision eruption timeline for the entire volcanic succession to show exactly when eruptions began and ended. Then, we would quantify the volatile releases from the magmas themselves and from the crustal rocks heated by or reacted with the lavas, to answer the question of the quantities and kinds of gases that were released into the atmosphere. Finally, given the timeline and the compositions, we could model the climatic consequences of these volatile releases. We had twenty-eight scientists from six countries participating. These people were specialists in experiments, in fieldwork, in modeling simulations, in the age dating of rocks, in atmospheric chemistry and climate change, in sedimentary geology, in eruption dynamics, and much more.

At last it was time to go. After our long day of travel from Boston and a quick night at the Bureaucrat's hotel in Moscow, we hauled our baggage to Vnukovo International Airport to board for Khatanga, a tiny town in north-central Siberia (we were surprised in flight to discover our seats were not bolted to the floor). We were watching and learning about our two new colleagues, Vladimir (Volodia) Pavlov of the Schmidt Institute of Physics of the Earth, a highly prestigious Russian Academy of Sciences center in Moscow, and Roman (Roma) Veselovskiy, soon to be faculty at the ultra-rigorous Moscow State University. They would be our companions and collaborators for at least the next five years. So far, I only knew that they were serious (серьезный), a stern compliment in Russia.

On that day we drove through town over roads that were packed gravel and coal dust, past a small empty lot where soccer might be played, and to a long, low wooden building known as the Geologists'

Hotel. Scientists stayed here while waiting for transit to or from the field. The building had a small entrance foyer, to contain the bitter winter temperatures, and then one long hallway with bedrooms off it, a couple of primitive bathrooms, and a kitchen, so we could cook meals. The building was unstaffed and freezing cold, requiring us to leave our coats on indoors. We settled in, because unfortunately another team had arrived just before us and commandeered the helicopters, and it would be some time—a day? two days?—before they were available to take us into the field.

The next day, Roma and Volodia went on their mysterious errands to assure our permissions and register everyone with the town. We had to lower ourselves into that utterly patient state of waiting, and of not understanding. This state was episodic but unavoidable in all the years of fieldwork we did in Russia. There were things that we Americans could not help with (registrations, permissions, negotiations) and there were things that we had to stay away from (intelligence personnel, bribes).

To pass the time, the rest of us walked to the riverfront and stood on the bluff, looking past some ruined buildings, perhaps still in use, and over the freezing width of river to the boggy Taimyr Peninsula, 250 miles wide, and beyond it, the Kara Sea and the islands of Severnaya Zemlya, the last discovered Arctic island, not sighted by Westerners until 1913. A chilly wind lifted my hair and I put on my knit cap. We were in a wild, wild place. Two hundred fifty-two million years previously, when the land was warmer but not otherwise much different, fissures had opened across this vastness, from slightly farther north than where we stood to almost as far south as Lake Baikal, over one thousand miles behind us.

From these fissures first flew fire fountains of ash and beads of incandescent magma, driven by superheated gases and fluids, but over time the eruptions calmed to flows like rippling cake batter. These

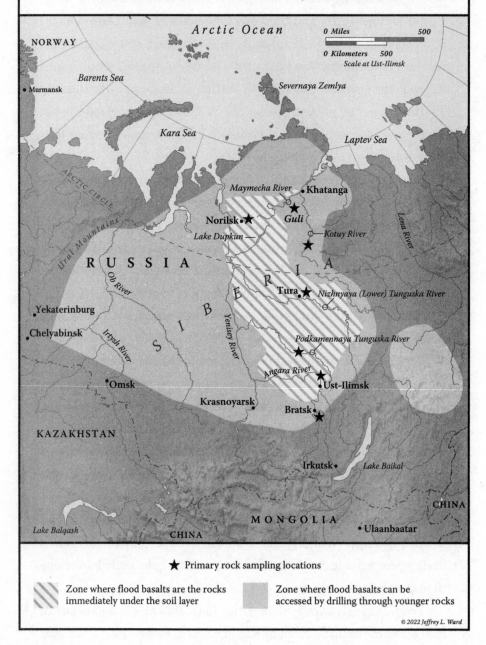

FIELD WORK IN SIBERIA, 2006–2012

Arctic Ocean

NORWAY

Barents Sea

• Murmansk

Severnaya Zemlya

Kara Sea

Laptev Sea

ARCTIC CIRCLE

Ural Mountains

R U S S I A

Maymecha River • Khatanga

Norilsk • ★ ★ Guli

Lake Dupkun —

○ — Kotuy River

★

Lena River

S I B E R I A

Ob River

Yekaterinburg

Chelyabinsk

Irtysh River

Tura • ★ Nizhnyaya (Lower) Tunguska River

○

Yenisey River

Podkamennaya Tunguska River

★

Angara River

★ Ust-Ilimsk

Omsk

Krasnoyarsk

Bratsk • ★

KAZAKHSTAN

Irkutsk • Lake Baikal

Lake Balqash

MONGOLIA

CHINA

CHINA

• Ulaanbaatar

0 Miles 500
0 Kilometers 500
Scale at Ust-Ilimsk

★ Primary rock sampling locations

Zone where flood basalts are the rocks immediately under the soil layer

Zone where flood basalts can be accessed by drilling through younger rocks

© 2022 Jeffrey L. Ward

eruptions continued, on and off, for over a million years. Siberia was gradually layered with lava flows in places as thick as seven kilometers from the Lena River on the east to the Ural Mountains on the west, from where we stood at the edge of the Taimyr Peninsula almost all the way south to Lake Baikal.

At the time they erupted, Siberia had just finished accumulating a vast amount of coal from rotted, packed, layered vegetation that built up over tens of millions of years. Under the coal lies as much as twenty kilometers of rock that formed when an even more ancient inland sea dried up, producing limestone, rock salt, gypsum, and the final crust of the drying sea, bittern salts filled with fluorine and bromine.

This natural progression of rocks adds up in the end to the perfect murder: heat it all up with magma, and the rocks sweat out and release into the atmosphere carbon dioxide, carbon monoxide, methane, sulfur dioxide, and halocarbons. Halocarbons are the class of chemicals including chlorofluorocarbons, the human-made chemicals that currently keep our air conditioners going and also destroy the ozone layer. No ozone layer would have irradiated all living things with ultraviolet solar radiation, and the greenhouse effect of the carbon dioxide would have held heat at the Earth's surface and caused temperatures to soar.

We suspected some of this, but we could not prove it without bringing back rocks to analyze. Ben and I were looking for lavas that contained the mineral olivine. Olivines are the first minerals to crystallize when magma begins to cool. As they crystallize, the olivine crystals trap tiny droplets of the magma inside themselves. These droplets, called melt inclusions, remain, frozen into glass, as the olivine crystals erupt with the remaining lava. Ben and I planned to measure the compositions of those melt inclusions, and thus learn the original compositions of the magmas before they erupted. As the magmas erupted, they would have released all the climate-changing gases into the Permian atmosphere. The only way we could know what gases were released was by finding those little pre-eruption melt inclusions, the time capsules of the past.

But where to look? How could we know which rocks would tell that story, covered as they almost all were now by swamps and forests? We had maps, beautiful, detailed geologic bedrock maps, telling us like a crystal ball what rocks lay under the swamps and along the rivers. In the middle of the twentieth century, as many as a million Soviets had supported or carried out summer field expeditions mapping the geologic wealth of Siberia. On April 1, Geologists' Day, the advance teams would march east over the spring snow to lay out food and equipment caches for the field geologists to follow as the snow melted. In early days these advance parties were trains of reindeer, and later, mechanical snow trains with caterpillar treads. Now the information from their numberless drill cores and meticulous field mapping was available to us all, allowing endless hypothesizing. The fact of Siberia itself, its vastness, its unreachability, its swamps and insects, had largely prevented further investigation by anyone but the most determined Russians.

We meandered around the town, stared at but avoided by the inhabitants. A pack of feral huskies, one with a grotesquely broken leg, trotted through weedy lots. A number of the houses had shops in the front rooms, each with shelves of items behind a counter and a solemn but friendly shopkeeper. With our rudimentary Russian we bought chocolate, and сушка (*sushka*), crunchy bread ring snacks like unsalted pretzels. One shop, in a purpose-built building, had a wet, dark foyer that smelled so strongly of rotting flesh that we retched and were grateful that the space was unlighted. The town was mainly silent beneath low dense clouds and showers of cold rain. Still no helicopter. Volodia brought us fresh hot pierogi, and tasted one hopefully, but then shook his head and decreed, "It can be eaten, but not with pleasure."

But our joy and anticipation of getting on the river was like electricity.

Finally, the helicopter was ready. I had a little irrepressible smile on my face, and I think the normally unsmiling Russians must have thought me an idiot, but the smile kept coming back. This was the moment we were going to be dropped in the wilderness. I felt a little rush of fellowship with Ben, when I saw he had that same foolish smile on his face.

We flew down the course of the wide, cold Котуй (Kotuy) River, the river we'd soon enough be paddling. We sat on bench seats, with no backs or belts, along the sides of the helicopter, with our baggage piled in the back and between us. I could see in the faces of my colleagues reflections of their experiences. Sam, with his extensive field experience and several trips in Russia, some of which included near-death accidents with Russian transport, looked serious. Brad, with his decades of experience canoeing the Canadian Arctic, looked serious and a bit more, maybe concerned, but he couldn't hide his excitement. Scottie, our cameraman for documenting the work, had the most adventure and exploration experience of all of us, and he was happy and relaxed. I came to learn that Scottie was happy and relaxed at all times.

From the sky we could see clearly the huge scars on each side of the river, where the ice had scraped all the vegetation and rocks clean for ten or twenty meters up from the waterline in its chaotic breakup and transit to the sea each spring. A herd of caribou trotted across a ridgeline, noses held high. And then the helicopter began to descend toward a point at the outlet of a smaller river, the кынлун (Kunlun), joining the Kotuy. We piled out onto the white Cambrian limestone blocks and unloaded our baggage in a rush and then the helicopter flew away.

We were here. After all these years, we were about to find the key rocks to explain the extinction, I felt sure. And for the first time in my life, I was far enough away from civilization that rescue was not at all assured. Even if my satellite phone worked, there was no guarantee that the helicopter in Khatanga would be there, or in working order, or able to make it up onto the plateau where we now were, since heavy weather often prevented its flying. An accident or illness or a bear attack here might well be fatal. For the first time in my privileged life, I could not just call 911, or phone a doctor, or shout out the window. It felt fantastic.

As we floated northward down the cold, powerful Kotuy, we would be floating upward through the geological ages of the layers of rocks, because of the tilt of the layers in this region. The layers of the rock rose from the river like an endless shelf of books, slumped at an angle. Layer after layer, rising up through time. We'd float up through the whole Tunguska sequence, and then we'd meet the flood basalts themselves.

The next day was full sun, clear as a bell, and the Arctic terns were flying in their Möbius bands and arcs over the river. Ben and I were interested in the "country rock," the sedimentary rocks that the magmas pushed into and the lavas erupted out of, and we were looking for lavas that would carry minerals with melt inclusions; most of all, we were looking for volcaniclastics, the results of explosive volcanic eruptions, not the calm lava flows that were more common. We wanted to see those explosive eruptive products. Sam was looking for rocks that might contain the mineral zircon, so he and his lab could determine the date of their eruption. Roma and Anya Fetisova, a graduate student, were there to collect small, perfectly chiseled cubes of lava, with their original orientation in the cliff meticulously recorded, to take back to the lab and tease out the direction and strength of the Earth's magnetic field from the time of their freezing. And Scottie and John, the director, were there to film it all.

Ben and I quickly worked out our sampling routine. The sample numbers all began with K08, because we were on the Kotuy River in 2008. Each unique outcrop was given its own number, in order from one, and each sample a decimal. The first sample from that year was K08–1, a piece of Cambrian limestone from our first camp. We wrapped tape around it and wrote its sample number on the tape with a Sharpie, and then put it in a heavy translucent plastic sample bag, taped that shut, and wrote the sample number on it as well. In our Rite in the Rain notebooks, we wrote the sample number, a description of the sample and the outcrop, perhaps a sketch of the outcrop, and then recorded an exact latitude and longitude. Sampling takes some time, and as the trip wears on, your baggage gets heavier and heavier with rocks. By the end of this trip, Ben and I would have collected 155 samples with a total weight of 180 kilograms.

I marveled at the frequent landslide scars on the steeper riverbanks; starting that night we heard their deafening crash as the permafrost that melted in the day gave way at last and huge mature trees and boulders fell from the tops of cliffs into the river. We learned to sleep on the shallow inside curves of the river, where there was no bank to collapse upon us.

I scarcely noticed when lunchtime passed by on that first day. Finally, after ten hours or so of travel and collection, we started scouting for a campsite. With the sun up all day and all night, it was really up to us to decide what our schedule was. We were all tired and hungry. Roma and I saw a likely looking beach, and we stopped and he and I jumped out to reconnoiter. Every inch of it was running with water. Water was bubbling in at the top of the cobble beach from the surrounding wetlands, and there was not any part of the beach that was not either streaming with water or had water trickling through the crevices beneath the top layer of rocks. Sleeping there would mean being saturated and, as a result, frozen. The nights were cold, and inside my heavy sleeping bag with its hood up and closed around my face, I was wearing a full set of long underwear, dry wool sleeping socks, a soft pullover, and a wool hat. Sometimes more. The next bend of the river was a couple of kilometers away. Between, steep rocky shores and wiry bushes meant pitching a tent was impossible. We had no choice, and we launched back out into the river.

When we found a decent spot, everyone set about pitching tents and making dinner. The Russians drank the water straight from the river, while the Americans shamefacedly filtered the water; eventually some of us gave up under peer pressure and started drinking it straight, too. Roma started a fire with driftwood. We propped a big round pot of water on stones and cooked macaroni (on other nights, rice, or *gretchka*) with ketchup and canned peas, corn, and meat mixed in it. We sat on rocks around the fire and ate from our bowls. We had tea with sweetened condensed milk (though the anaerobic mold on the insides of the lids concerned me, earning Anya's disparagement), and some little hard cookies. But the tone was more silent than celebratory.

Once we'd gotten some hot supper into our stomachs normal conversation was possible once again. With embarrassment, I asked Roma about lunch. "Lunch?" he asked. Apparently this was not done while working. The schedule was to get up at a comfortable hour, eat a hot breakfast, and then work ten or twelve hours. Camp, dinner, write field notes, sleep. In conflict, the deeply ingrained North American field practice involved getting up at dawn, having a hearty lunch partway through the day, and camping in time for a little relaxation before bed. Both these schedules are efficient and enable a lot of work. Field time is precious; you may never see that wild place again, have another chance to sample that key rock. In my mind I saw both Sam and Roma as senior dogs with their hackles raised, each sure their own way was the right one.

I reasoned with Sam, off on our own, that with twenty-four hours of daylight, who even knew what getting up early meant? And couldn't he write up his field notes over coffee while waiting for Russian breakfast to commence?

And I asked Roma and Anya if there was some way we could eat a small meal in the middle of the day. Thus began the hilarious morning routine of Anya straight-facedly handing each of us Americans a Snickers bar to carry with us and eat for lunch when we needed it: "Leenda, Sneekers. Sneekers, Brad." For Brad, this was particularly ironic, since an allergy prevented him from eating them at all. The giant bag of Clif Bars I'd brought suddenly took on a starvation-level value. I gave Brad enough for his lunches. The rest of us savored our Snickers. Years later, as I battled ovarian cancer, Sam sent me a bag of Snickers. Thinking of the poetry of that gift in a moment of terrible desperation in my life makes tears slide down my cheeks.

The challenges of binding together this team continued. Anya came back from her first big day of collecting with tear streaks down her own cheeks; she'd had to climb out on cliffs in ways both unfamiliar and frightening, and chip out perfect samples, or else start over. We climbed those cliffs with no safety gear, no helmets. Not all of us were natural mountain goats, and it took some nerve. Anya was building hers in real time. Between that tension, the late starts, and the Snickers, let alone the nervous-making boats, there were a lot of side conversations for me.

Disagreements among team members were more concerning when they were about the rocks. I was thrilled to find beds of ashy volcaniclastics with boulders mixed through. Sam dismissed them as later reworkings of earlier eruption products, that is, that they were not the originally laid-down materials. I did not see how that lessened

their importance: these rocks showed there *were* ashes and volcaniclastics, explosive products, reworked or not. We did finally agree, on
that river, that those rocks carried the evidence for ashfall, evidence
for rainfall and lahars, and evidence for debris flows, suggesting a violent and explosive environment like the eruption of Mount Pinatubo
in the Philippines, not a calm-flowing lava like that at Kīlauea.

At one point I was thrilled to see a thick pale rock unit high on a
cliff that I was sure was a tuff, a fine-grained volcaniclastic rock. Sam
said no it was not. Sam was far more experienced in the field than
I, so arguing back about a rock unreachably high on a cliff was a long
shot. Usually, we happily discussed the provenance, eruption patterns,
the minerals in the lavas, and everything else about the province. But
these mostly silent disagreements when my opinions were flattened
were tough to take. The following year, Ben got up to that high unit to
take a sample. It was indeed a tuff.

But I was entirely contented. This was an adventure, and we were
on it. Ben, bless him, had no problems. In the middle of our long days
of sampling we played that classic field game: slam your field notebook
shut, and whoever crushes the most mosquitoes wins.

After spectacular days of sampling, we had a long run, twenty-five
miles as the crow flies but who knows how long as the river winds, to
get to tiny Kayak, home of one hundred coal miners. At one sampling
stop along the way, lucky outcrop #13, the cover on one of the pon-
toons split, and the rubber inner tube bulged out like some revolting
giant intestine. We bound it up with duct tape as best we could. If
that pontoon blew, all our food, most of our samples, and several of
us would be thrown into this deep, swift, freezing river. I watched
the scenery go by, and watched the pontoon, and tried to estimate
whether we were listing more than we had been. We finally limped
into Kayak around seven in the evening, and hauled out on the town
beach. Even John's inflatable boat was deflating faster than usual, and

as he pulled up to the beach, he just looked at me from his dolorous eyes.

We walked out to find some food in Kayak. We found a little shop run by Sveta and Lyuba. We bought groceries—a pig's foot, peas, rotted potatoes that we were given for free, vodka, sardines, an onion, chocolate, beer, and tomato paste—for 1,200 rubles or about $60. When the rest of the team had moved on down the street and I lingered behind, Sveta and Lyuba brought out their homemade currant wine, and the three of us told each other about our children, and they advised me that fieldwork was no job for a woman, and I should go home to my family. I felt that familiar, disorienting oscillation of belonging/not belonging: One moment I was a woman talking with women, and the next they told me I wasn't acting as a woman, and couldn't be a field geologist, because that was for men. Not woman, not man. What was my role in this world?

Sveta and Lyuba also wanted to know: Why Russia? I got this question often. Just the fact that the flood basalts were there didn't seem to be enough. I had two reasons: First, Turner's great uncle had been Chip Bohlen, the American ambassador to the Soviet Union under Khrushchev. That statement was almost invariably met with stony-faced silence.

Second, I said, my father's family was Russian. They had come to the United States around 1917, when my grandfather was a teenager. On his deathbed, my grandfather told me stories of the terror of the Cossacks, and of life in little Rogachev, near the Dnieper River in current Belarus. Almost invariably, I'd be met with smiles and patting hands on my shoulder and questions about their names. Suddenly, I felt more Russian. And then, wonder of wonders, one day I saw some Russians putting fruit jam into their tea. I almost fell off my chair. My father used to put fruit jam in his drinking water and into his tea. I had always thought it was just another of his inexplicable eccentricities.

But no, it was pure Russian, come to him honestly from his own father. I suddenly felt my roots as more than theoretical.

THAT NIGHT, IN ONE OF THE TINY BEDROOMS OF THE GUESTHOUSE, Roma and I settled down to have a cup of tea and discuss the day. We could hear drinking parties in other rooms. In the hallway on the way to the communal kitchen, though, a small, strong Russian man with little language of any kind displayed an alarming wish to grab people, particularly me. His wife hit him on the head and he stopped. Today was the coal miner's first payday in five months, we were told. The next day, Sunday, was utterly silent throughout the town.

Beyond that split pontoon, I was also limping into Kayak in another way. My knees were causing problems again. I had built up my quadriceps muscles in advance of the trip, but one misstep had, I found out later, torn my meniscus. At the time, I only knew my knee had swollen like a melon and in just forty-eight hours my quadriceps had atrophied as the nerves shut them down, protecting the damaged knee. Scottie helped me wrap it strongly. I'd been plagued by knee injuries my whole life and this was deeply frustrating. I decided that in Kayak I got one day of rest while the others took a boat downriver to a certain outcrop, and then we'd just keep going.

Sunday morning the team went off in a rented powerboat to sample the outcrop downstream, and I went out by myself to find the банная (banya), the public sauna and bathing facility. Wildflowers waved along the sides of the town's few dirt streets and the sun was shining. As I went along the boardwalks toward the banya, I saw the deranged man from last night's hallway walking along, about a block behind me. There were no other people in the streets. I kept my eye on him. He was walking a bit faster than I was, gaining on me. I walked faster. I remembered the bizarre strength of his arms as he grasped at me last night, and his wife hitting him over the head. I reached the *banya* and ducked inside, and quickly paid the attendant behind her half door. The man came in the street door, and I hurried into the women's bath and locked and barred the door behind me, my heart beating fast. I was not confident that the attendant would help, if I were in trouble. As I stood astonished a couple of meters from the door, I heard the man outside scratching the door with his fingernails. I waited, silent, ready to fight. He did not say a word, but he scratched for a long time. Finally, the scratching stopped.

I had a lovely sauna and shower and massaged my leg and knee. With some trepidation I opened the door when I was finished, but the man was gone.

OUR NEXT STOP MIGHT HAVE BEEN ONE OF THE MOST EXOTIC GEO-logic locales in all of Siberia: the Guli geological province. I think we were the first non-Russians ever to go there. The Guli province was discovered, we were told, in the 1920s by prospectors sent out on reindeer-back from Norilsk, about 350 miles away. They arrived, and built cabins, and hacked through the shattered frost layer with pick-axes, uncovering geological treasures. Guli was likely a huge magma chamber that slowly crystallized, and was now eroded away and revealed at the surface. As the magma froze, first the mineral olivine solidified and sank to the bottom of the chamber. Olivine has no nutrients and some toxicity for plants. A ring of this rock about thirty miles wide exposed on the surface had almost no surface vegetation. As crystallization continued, the remaining magmas had stranger and stranger compositions, as all the common minerals solidified out. Finally, the magmas were producing ore minerals, and emeralds, and gold, and masses of shining mica, and the last dregs of the magma were almost pure calcium carbonate, a very rare kind of lava called carbonatite. We wanted to sample to see if we could identify the source of the original magma. Did it melt from the mantle, as the rest of the magmas seem to have done? Or did it have contributions from the crustal rocks? (For the record, we did not see any emeralds or gold while we were there. We were sampling economically unimportant but geologically fascinating rocks.)

The helicopter came down on flat ground next to the tiny Guli River. We found what looked like boxcars there on the tundra. They were indeed train cars, but from a snow train, not a normal train—no train tracks or roads come anywhere near the Guli. These snow trains have cars on great skids, like skis, and are drawn hundreds of kilometers across the winter landscape by big caterpillar-treaded crawlers. The last people to mine at the Guli had left these behind. One of the snow train cars had beds, and an indoor space complete with a calendar from 2001 where we could sit to eat, out of the weather. Another was a sauna car, which Roma made functional one evening.

Some slopes were just weathered rock and were lovely for walking, but flatter areas were covered with lichen and moss, which is like walking across piles of firm round wet sponges. The wildflowers were as varied, delicate, and delightful as they are in spring and summer in the Alps. We saw a huge arctic hare, big as a red fox, and lots of wolf and bear scat. We came through the old settlement from the 1920s and '30s, a small group of log cabins, which we peered into. We walked through a couple of them, seeing the abandoned wooden furniture, the bunk beds, and some mica crystals as big as dinner plates that they had left behind. The loneliness of the snow train cars seemed all the lonelier for this other, earlier abandoned settlement a few miles away. These wooden cabins felt like homes.

I was especially excited to see some rare rocks that I had studied, remotely, for years. In just a few locations volcanoes have erupted lavas exceptionally high in magnesium, and low in iron and silica. Their strange composition indicates that they melted unusually deep in the Earth, and at unusually high temperatures. Their very existence here strongly indicated that the flood basalts had been formed from a plume of solid mantle rock, rising from at or near the core-mantle boundary, and unusually hot. Because of its excess heat, the plume material would melt at a higher pressure (greater depth) than more temperate mantle material. The high temperature and pressure conditions meant a slightly different composition of melt would be made: these very high-magnesian, low-iron and -silicate lavas, called meimechites.

A few years previously, at Brown, I had done experiments on tiny artificial samples of meimechites held at a range of temperatures and pressures, similar to the way I had worked on the feldspars during my master's thesis. These, though, required far hotter temperatures, up to 1,800 degrees Celsius, and far greater pressures, up to 7 GPa, about the pressure that would be bearing down on you if you were buried 150 miles deep in the Earth. These experiments showed that the meimechites had formed at about 5.5 GPa, the equivalent of about 110 miles' depth in the Earth, and 1,550 degrees Celsius. These are extreme conditions for melting inside the Earth. Most melting, under arc volcanoes or at the mid-ocean ridges, for example, occurs around 1,300 to 1,400 degrees Celsius, and at a depth of less than 60 miles. These meimechites, then, are direct evidence for the origin of the Siberian flood basalts being a hot mantle plume from deep, near the core.

And that day in July 2008, I was actually going to see meimechites in the field, in this wild, inaccessible central-Siberian taiga. Roma, Anya, Ben, Brad, and I looked together at our maps, and set our GPS devices, and started off. The low clouds stretched seemingly to infinity across the long low country, and we wore our rain shells and wool hats against the chill. We walked up a long, slow slope, across the brow of a hill, and down another long, slow slope, until we came to a river.

Crossing a river was always exciting. The fastest method was to leap from rock to rock, when they were available. But I'm not especially agile, certainly not with that injured knee, so I generally chose to wade and get my feet wet instead. We pushed through a willow thicket so dense that we could not see each other when we were just a meter apart. By the third of these hill-river-thicket combinations, we'd walked perhaps five miles, and we were getting tired. We trudged slowly up another long, long shallow hill, the footing again like walking on wet sponges.

But at the top were the meimechites, dark green knobby outcrops lying along the top of the hill like a Mohawk. The ground was carpeted with grasses and little arctic flowers that looked like cotton balls, and the views went on for miles over rolling hills, sparse larches, winding rivers, distant rock outcrops streaked with snow. The thrill kept a smile on my face; I could not stop exclaiming as we found wonder after wonder: rock faces thick with olivine crystals, or crazed with a grid of bright green asbestos or black magnetite, from fluid fluxing through the solid rock and reacting into new minerals. We took photographs, using for scale a stray caribou tooth from one of the several skulls or jaws we found. We walked both sides and the top of the outcrop, photographing it all, and taking samples with difficulty, because the rock was very hard. The small sledge and cold chisel were needed. At my first sample, the satellite GPS read N70.86433°, E 101.21465° (throughout these field seasons, we were to sample from 57.9°N all the way to 72.5°N).

In sampling, as in all aspects of fieldwork, it seemed, I had the disadvantage of being less strong than the men. Anya suffered from the same disadvantage, but she came at it from a different position in the team. She and Roma were taking samples for a different purpose, and she was a junior member at that time. She was a beginning graduate student, and in the field with a senior scientist. She was twenty years younger than I, tall, blond, and determined, and she was living through a different period of having to struggle to be seen and treated as worthy. I, on the other hand, was just becoming conscious of another moment of surprise at not being quite welcome.

Anya and I were always welcome to cook. But I didn't want to cook. I wanted to carry my share of the load, to take my own samples, to chop wood and build fires. But I had to fight—delicately, gently—for those rights. At times, I'd be working to get the chisel in just the right place and then to give it a big whack with the hammer, and I could practically smell the silent impatience from the men nearby. Yes, they could have done it faster, and with fewer blows. But why should that be the important metric? Why is it not more important to let each person do the tasks they want and need to do, at their own pace? So much of field experience, for me, was waiting to be criticized, or consciously not caring if I was criticized. I often thought to myself, in some cultures I was as big and as strong as most of the men. So why was the fact of my being smaller and less strong than most of my Russian and American male colleagues a constant tiny undertone of discontent, and even sometimes voiced as a burden to the team, or a failure of some nebulous sort?

So often I had the tool taken from my hand by the man who wanted to do it faster, or to show me how to do it. Then the expectation of interruption, even subconsciously, led to a slight hesitation whenever there was a physical thing we had to do. That slight hesitation in action created that little vacuum, the same as in a conversation, and often a man would jump in and start the task. It became a form of learned helplessness. Maybe I really couldn't do it, I began to think. Certainly, I had less practice than many of them did. Maybe I should just be doing something else.

I noticed that I really enjoyed doing things when I was alone. Alone, there was no one to imply that I wasn't strong enough or fast enough, or that I spent too much time thinking before I acted. I could relax and experiment and find ways to succeed. And then I noticed that working with certain people, male and female, felt just as good as working by myself. Just the courtesy of their asking a question instead of stating a contradiction left me enough space to keep owning my progress. By the time I came to Siberia, I'd built up a lot of physical expertise on my own: I'd built my own bicycle, I'd made those high-pressure furnaces with Tim, I'd designed and built my own high-tensile electric fence around my own sheep pasture. But the struggle to hold the tool was not over.

Part of the struggle, I'd always known, was in my own head. If I walked out there determined and courageous and just did it, normally no one would stop me. But that conscious daring is tiring. In Russia, I had to contend with explicit bias against women, not only the normal implicit bias I experienced where I lived and worked in American academia. Russian men seemingly were not used to women in scientific leadership and many were uncomfortable with it. Russian women consistently advised me to stop and go home. And then, as I got to know them, Russian women scientists began to confide in me. They asked advice for how to get ahead; they said that they were still expected to do all the housework and child care, and also, that domestic violence was widely tolerated.

On these Russian trips I was able to define, in the end, a region of work that could be mine without intervention: taking my own samples, carrying my own baggage. To do more, I had to insist, but also to try to insist gently and with quiet good humor so that my participation was not awkward or disruptive. I was almost always, for example, shouldered aside from the line of men carrying all our team's baggage from a guesthouse to a truck. Sometimes, I could not tolerate standing by and watching the camaraderie of that work, and I would have to join. Never did that go unremarked by our Russian colleagues. I don't think I ever really got the balance right.

When at last we had photographed and sampled and admired the last of the meimechites, Brad, Ben, and I walked back to the other end of the outcrops to find Roma and Anya. They had finished sampling and had made us hot soup for lunch! We gratefully accepted bowls of soup from the pot over the fire, and we laughed at the numbers of mosquitoes that flew to the heat and drowned, even in the very spoons of soup on the way to our mouths. But then, we thought to ask, "Did you walk all the long way back down the hill to the river for the soup water? Wow, thanks!" But no, Roma and Anya laughed, and said, "We found that little pothole in the rock right over there and ladled the water out." We looked over. The little pothole, the size of a kitchen sink, had plenty of caribou droppings floating in it. Ah, well, we thought, it's been boiled! The soup was delicious. And then we hiked all the way back to our camp, through the three rivers and willow thickets and miraculously freshly green Siberian larches.

We learned from Anya that it was Roma's birthday. Anya made a plate of sweet no-bake cookies, and we saw more clearly with every moment how deeply she felt for Roma, and, I thought, he for her, though he gave less away. We loaded in our baggage (I managed to carry my own) and jumped into the helicopter when it came for us, and soon we were on our smooth ascent and I was inwardly already mourning that amazing place, a place so hard to get to, a place I would never see again. Until a trip is over, it's hard to know with what emotions it will be remembered, whether it will feel like success or failure. I knew then that we'd be friends for a long time.

The next day we went to the airport and packed our gear and samples into a shipping container, and waited in the little wooden airport for our plane. There was a poster on the wall that Roma and Ben and I translated with the help of a dictionary: It offered cures for weeping, hysteria, nervous trembling, hyperactivity, aggression, stupor, and apathy. Many nights in the field Ben and I had been taught some Russian by Roma, and had given him some new English words in return. As a pointed lesson, one early night on the Kotuy, Anya taught me the words she thought we Americans liked best: сейчас, здесь, быстрее (*now, here, faster*). Here in the airport on our way out, Ben was asking Roma for more words, but then he interrupted himself, laughing, to say he had all the words in Russian he could possibly need: "Beetle . . . Breakfast . . . Marry me. . . ." And I was reminded of the twenty years I had on these good young men.

The next day we flew on first to Khatanga (and on the way, in our cargo plane, where we sat on the steel floor as if we were camping, we discovered that the whole back of the plane was filled with frozen, flayed carcasses of caribou, stacked up like cordwood and beginning to smell) and then to Moscow, had a night at the Bureaucrat's hotel, and at last, hopped to Paris Charles de Gaulle Airport on our way back across the Atlantic. As always, back in the Bureaucrat's hotel, I had showered three times, four times, trying to wash the grease and sunscreen and bug dope out of my hair, trying to produce also some clothes clean enough for a long flight home. Coming out of the Jetway in Paris we were hit by a wave of perfume from the shops, and blinded by the brilliant lights of glamour commerce. Suddenly, whether it was raining on us, what we had to eat, and where I could find to go to the bathroom, were no longer the top of the mental list. But I was filled with hope that the 850 pounds of samples we'd taken would yield the ages of the eruptions, and we'd be on our way to answers at last. I did not yet know that though Ben and I would be able to extract much of the story of gas release, atmospheric chemistry, and climate change from the rocks we collected that summer, Sam and his graduate students would still not have the samples they needed for dates.

Chapter 6

PAST IS PROLOGUE

Partway back to Ust-Ilimsk on the Angara River, motoring south, after sixty miles of pristine wilderness, Sergei invites us to spend the night at his fishing camp at the mouth of the Kata River. Sergei's fishing camp consists of several nicely built wooden houses with a collection of outbuildings, all connected by neat wooden pathways and set into a great meadow, mowed into lawn near the buildings. At the dock we are met by two young Siberian huskies, Pulya and Chama, bowing to Sergei, wriggling, yodeling. I immediately acutely miss our border collies at home, and Turner and James, and I even miss a Siberian husky from our past, Natasha. Dogs.

There is a bear skin on the wall, and there are great soft couches, and a satellite television, and a framed photo of Putin bare-chested on a white horse. Volodia makes macaroni with meat mixed in and we add hot sauce and horseradish. We have a little pepper vodka, and some tea. Then Sergei tells us the sad news: a big hydroelectric dam is being built a distance downstream, north of here. The dam is being completed this year, while we are there on the river, and as soon as the dam is complete the river will start to rise. The water is expected to rise twenty yards at the Kata mouth, where we are; the whole fishing camp will be destroyed and submerged underwater. Downriver, authorities have already evacuated and burned to the ground a whole village.

AFTER THREE FIELD SEASONS, WE WERE STILL MISSING THE UNDENI-able evidence for hot explosive volcanism that could have carried these toxic and greenhouse gases up into the stratosphere, where they would circulate around the globe. None of the samples we'd laboriously col-lected and carried home had yielded the minerals needed to make the sensitive age measurements. Then we found a short paper in Russian about a study of ancient pollens. The authors used samples from the Angara River, where they said the rock outcrops in a certain region were all volcaniclastics: all rocks formed of fragmented minerals and glasses produced by explosive eruptions.

And so we had our 2010 field plan. We were off to the Angara River, starting with a train trip to the town of Ust-Ilimsk. There, we checked into a hotel and began making arrangements for the river trip to come. In the couple of days we spent in Ust-Ilimsk, I had more time to talk with Volodia.

He told me more about his parents and his childhood, and we com-pared notes on taking care of the aging and the ailing. He also shared the unsurprising news that the man who had come to my workshop at Brown back when I was just starting the project, and had taken ideas away and never collaborated with us again, was jealous of our funding and progress. We had hoped he would allow someone on one of his trips to collect a few samples for us, but he refused. I felt discouraged that these small problems and disagreements had come up, despite my best efforts. And late one afternoon, Volodia and I sat down to have a mug of tea together, and he put his forehead on his palm, shaking his head, saying with such relief that the man whose boat we were going on was a good, sober man who would keep us safe. In a flash I could see into the stresses of what Volodia managed for all of us: trying to find a boat and a captain from far away in Moscow, not knowing whether he would be honest, or sober, or safe, until we were all already there. I shared his relief, and felt a weight come off me, too.

We boarded the *Grom,* a steel-hulled ship just big enough for us all. The cabin wouldn't quite fit us all comfortably on its bench seats, but the wide green-painted deck was a fine place to sit and watch the river cliffs go by. The boat had a terse and level-eyed captain, Sergei, and a tall, sinewy, tanned, and silent boatman named Oleg.

After motoring for a couple of hours, we saw the first outcrop. It was clearly the result of explosive volcanism. And it reached from the waterline up to the top of a high cliff, only an erosive surface at the top, perhaps ninety yards of volcaniclastic visible with more below the waterline and more gone forever from the top. I almost could not contain my excitement. We were all on deck, staring in amazement at the cliffs, and Scottie was photographing everything, us, the rocks, the water . . . The next outcrop was a volcaniclastic. And the next. And every one for 150 miles. Here was one of the largest regions of explosive volcanic rocks known in the geologic record. The Siberian flood basalts were preceded by immense volcanic explosions, sufficient to drive climate-changing gases into the Earth's stratosphere. Here they all are.

We would be the first to study and write in detail about these results of explosive volcanism. And beyond the excitement of finding them at last, they turned out to have inside them the microscopic remnants of the biggest coal-burning event prior to the industrialized age. After we motored 150 miles, we turned south again and began sampling, and camped on the riversides and slept in our tents, and motored along, all the way back south toward Ust-Ilimsk.

And now, at the Kata River confluence, where Sergei's fishing camp is, we find the rocks we are most anticipating. As reported in that scientific paper about pollens, they contain lapilli. Lapilli are spherical objects, about the size of a marble, which are formed in the roiling hot ash clouds of an explosive eruption column. Some small initial particle is coated again and again by layers of material. This process is the same that makes a hailstone in a thundercloud, but in the case of lapilli, the layers are made of tiny mineral and glass fragments ("ash") rather than ice. The lapilli are abundant. And here, as at many other Angara outcrops, I find black pieces of charcoal. I begin to wonder about this charcoal: Is it plant matter from burning, or could it be pieces of coal? Surely after the surface is covered with fifty meters of boiling-hot ash, there could be no more trees to burn, and anything else must be coal? But the answer to this question would have to wait a few years. For now, I collect samples and record them in my field notebook, but I focus on the lapilli and the tuffs.

At the outcrop early in the morning the grass is filled with newly hatched white Parnassius butterflies, still too cold to fly. In the afternoon Ben and I see skinks, and bright orange caterpillars, and eagles of some sort. Ben shares my intense interest in all the living things around us, and is a butterfly expert. The soft reeds by the camp are filled with ducks. The sun is setting over the mouth of the Kata, its reflection shining yellow and gold in the water. We are collecting some of the most interesting and important rocks of our lifetime. Ben and I exchange looks. In one year, because of the new hydroelectric dam upstream, the outcrops will all be underwater, effectively forever.

The teamwork on this trip is better than in 2008. We have more productive conversations about the meaning of the rocks, and the possible paleoenvironments in which they formed. But still, I write in my field notebook: "It's a good team. My only difficulty seems to be in the perennial problem of being condescended to, not least by Sam on the outcrop. He has much more experience than I do but he is not infallible; it's hard to be contradicted with such ironclad confidence." The damage of flat contradiction is not just to me personally, but also to my scientific reputation in front of colleagues and particularly graduate students. I have watched graduate students, particularly men, learn the practice of harsh contradiction instead of discussion, and I've watched them begin to practice it on each other, and on female faculty. This practice does not indicate the depth of the person's knowledge but is instead a mark of the senior academic in one kind of traditional style; it's a way of saying, *I am master of my field*. This manner is a reason for some excellent junior scientists leaving the field. And this practice does not lead to best learning and discovery. Sometimes I have been wrong, and sometimes they have been wrong, but the wrongness is not discovered until later, and the lessons that could have been learned from it are thus lost.

I begin my ongoing struggle with how to portray myself as knowledgeable and senior, without driving over other people, and without occluding complete conversation of everyone's ideas. Asking questions, real questions, rather than offering comments or criticisms disguised as questions, is one really good way to bring everyone into the conversation.

We headed home from that field season elated, and laden with 150 pounds of samples. Ben went into the lab and searched for evidence of the temperatures at which these rocks erupted, which would predict the height of the eruption plumes and thus whether the gases they carried would be injected deeply into the atmosphere. Seth took his samples to Sam's lab and began searching them for the mineral zircon, which would give him the most accurate dates for the rocks. And I headed back to work on some theory of planetesimals, those earliest small planets of the solar system, the building blocks of what we have today. I'd continued the magma ocean work into a smaller scale for earlier objects in the solar system, and the work was beginning to get interesting.

IN 2011, I MOVED MYSELF AND MY LAB FROM MIT TO THE CARNEGIE Institution for Science in Washington, D.C. To my great surprise, their search committee had invited me to be considered for a directorship, and to my greater surprise, they made me the offer. Maria Zuber, the head of my department at MIT, and Marc Kastner, the dean of science, and even Susan Hockfield, MIT president, met with me and kindly encouraged me to stay, offering to expedite tenure to that year, and support me in other ways. But it was not a question of support; MIT still felt like home for me, but the lure of the adventure and the leadership position were irresistible. And so when I left for our final season in Siberia in 2012, it was from Washington rather than Boston. By now, despite the different departure city, this all felt normal and easy; our team's gear was familiar, and our habits, and the foods we each liked, and we had an arsenal of jokes from previous years.

Tura, a town of about five thousand, used to be the capital of the Evenki Autonomous Zone. At its height, the town had about 25,000 people, but they lost a power struggle in 2007 and their zone was merged into the larger Krasnoyarsk Krai region, and they were no longer the regional capital of the past. Even in 2012, though, the town was an interesting mixture of Evenks, Yakuts, and Kets, all First Peoples, and Russians, Ukrainians, and others from outside Russia. Tura felt small and isolated and a bit too sparse; the fall-off of population was still evident. We were told that the only remaining paying jobs were government jobs: road crews, teachers, medics, and the like. Everyone else was a subsistence fisher or hunter. There was talk of moving the whole town about one hundred miles south to join another settlement.

By 8:30 A.M. on July 14 we were on the river in a fine steel-hulled boat that belonged to the local rescue force. Why this boat was at our disposal is knowledge above my pay grade. The boat came with Captain Piotr Nikolayevich, mate Dima, Piotr's young son Artiom, and a yellow husky named Malashok. We all were charmed by smiling, shy Artiom, and friendly, lazy old Malashok. That first day on the river I sat for hours on the bow of the boat, anorak hood up against the cool air, gently patting Malashok, who lay on his side by me. As on the Angara, we went to the far end of our planned sampling region on the river, and then started back, taking the samples on the way home.

That first night I was able to pitch my tent on sand; what a relief to sleep on sand instead of rocks! But the night was very cold. I put on all the clothing I had: long underwear, woolly yoga pants, two pairs of dry thick wool socks, extra fleece pullovers, a neck warmer, and a thick hat, but my buttocks and thighs had gotten chilled before I went to bed and having them in my mummy bag with the rest of me was like snuggling up with a block of ice. I'd been careless just standing around or sitting, waiting. Warming again after a deep chill is difficult, and we had no prospect of a warm place ahead and no long hikes, so I had to work hard to warm up in the morning. About an hour after a завтрак (*breakfast*) of cornflakes, yogurt, and cocoa-flavored sweetened condensed milk, off we went on the boat. My Garmin GPS had eleven satellites located and read N 64° 9.136', E 101° 19.303'. We were in the middle of the endless Siberian taiga, the trees thick and tall, crowding the river.

We were getting spoiled with great rocks. Just like on the Angara, all the river outcrops along the Nizhnyaya Tunguska for the over 150 or so miles that we traversed were tuffs and other kinds of volcaniclastics. The region where the first Siberian flood basalt eruptions were explosive was vast. And here also were pieces of charcoal, as well as chunks of coal and sedimentary rocks that had been carried up from depth by the force of the eruptions.

Imagining this landscape at the time of eruption requires a kind of cartoon mind, because we have nothing like it on Earth today. Vents would open and spew out fragmented flakes of minerals and glass (the ash) driven in plumes of superheated gas, including water, and carbon dioxide, and halocarbons. Huge plumes would form and rise into the stratosphere. They must have roared more loudly than rockets. Elsewhere, liquid magma would ooze out and flash lakes into steam. Whatever vegetation had taken hold since the last eruptions would be burned and leveled. Precarious steep-sided cones of ash and rock would form around vents, only to avalanche down in rain- and tremor-induced landslides. This, as far as the eye can see, farther even—walking across this region, if you could without being killed, would take weeks.

On our last afternoon on the river we passed two men in worn, dark clothes walking along the river's edge. Our captain put on a little siren and pulled the boat over to them. I asked why the captain did this, and Volodia explained that because the men had no boat, no tent, and no packs, the assumption was that they were in trouble. Indeed they were. These two small, bent Evenks, looking like wild Asian mountain men, were walking to the "hospital," a little clinic in the settlement of Nidim. One of the men was very sick. We took the sick man onboard, and the other turned and walked back into the forest and disappeared.

They had been counting on meeting a boat, because it was at least another thirty miles to Nidim. The man said his liver was bad. My colleagues said this was probably from drinking too much bad mineral spirits, and I thought back to the threatening man in Kayak, who had been drinking mineral spirits in the guesthouse, and who stalked me through the town. This sick man sat silent on the hatch cover. He answered a couple of question for the captain, and accepted a cup of hot tea. He then lay unmoving on his back for several hours, and later sat up looking more clear-eyed, but when spoken to, he stared back blank-faced and said nothing. We were much quieter than usual, sitting isolated around the deck. Periodically Dima checked to make sure he was still alive. Eight hours later we arrived in Nidim, and he slid off the boat and walked up to the clinic to die; of course there was no liver transplant possible. As Piotr and Dima remarked, the rescue squad's primary business was picking up dead bodies.

BACK IN KRASNOYARSK, ROMA AND VOLODIA BROKE THE BAD NEWS: our permissions to work in Norilsk had not come through. All the places we went for fieldwork required layers of permissions, but Norilsk was particularly protected. Norilsk is one of Russia's "closed" cities, presumably because of the potent mix of vast mining operations and missile silos. Only a couple of hundred foreigners were allowed in Norilsk per year. In answer to my every question about plans, our colleague Anton would answer, "It is not known. It cannot be known." Seth, a student named Alex, and I stayed in Krasnoyarsk, and Roma, Volodia, and Anton went on to Norilsk to try to sort out the permissions. The alternative for us was to fly to Norilsk with them and then spend three days in jail.

Krasnoyarsk, the first year we visited, had very little aside from one large, dark restaurant near the hotel. Each subsequent year we went, unlike the smaller Siberian towns and villages, Krasnoyarsk looked more affluent. This time, there was a stylish modern coffeehouse with a brushed-nickel front door, and a fancy steak house, and there were Bentleys being driven down the roads. The oligarchs had arrived. We walked all around the city during the day, looking at the beautiful old carved wooden buildings, and the new flashy additions, and drinking квас (*kvas*, a mildly alcoholic drink like beer) and молоко (*milk*) from miniature tankers that vendors pulled along the sidewalk on wheels, and decanted drinks from a spigot into glasses the waitstaff took back and washed and reused for the next customer.

Four days later, we were legally in Norilsk. Through the patience and persistence of Roma and Volodia, we were able to travel for our field seasons to all the places we hoped to. And so we resumed our search for zircons. Seth had spent three years processing every likely sample we brought home, and had found almost no zircons. All the zircons in the volcaniclastics were inherited, that is, they were from the sedimentary rocks of the basin, not the magmas, so their ages predated the flood basalts and the volcaniclastics. In general, zircons crystallize from more "evolved" magmas than the primitive, straight-from-the-mantle basalts of the volcaniclastics (all basaltic in composition) and the flood basalts. To evolve, the magmas need to cool and partially crystallize. As crystals form, the remaining magma's composition changes. Elements like silicon and zirconium are enriched in the remaining magma because they are less incorporated into the early-forming minerals. Finally, the remaining magma has so much silicon and zirconium that the mineral zircon is stable and begins to crystallize. We had had no luck finding any evolved rocks. We were thus worried for two reasons: first, because we needed to nail down the age of the eruptions far more accurately than had ever been done before, and Seth had worked out the near-miraculous new lab procedures to do it, but we needed the rocks; and second, because Seth wanted to actually produce results so he could graduate with his Ph.D.

We went to the famous красная камни, or Red Stones, a waterfall and gorge up a hillside carved into rusty red lavas from the flood basalts. There, we found one sample that really looked like an evolved magma, in a vein suggesting it was the last liquid of the flow, but back in the lab over months and years it would come to nothing. We also saw there the charcoalized stump of a tree that had been overrun by the lavas, 252 million years before. I saw forget-me-nots, one of my favorite wildflowers, growing there. Volodia told me they are called forget-me-not in Russian, too.

We had dinner again with the wonderful Viktor Radko, he of the serpentinite box, and the black bread toast to Mother Russia, and of taking his cat into the field with him in summers. "His cat has moved to Belarus, he has gone on *vacances*," explained Volodia, translating to English with a little bit of French. And then a moment later, "Oh no, I misunderstood, he is dead." We then learned the word мерзко, *worse than awful, cursed*. But then, it turned out, the cat really *was* on vacation in Belarus.

THE NEXT DAY, WE HAD ONE OF THE MOST MAGICAL FIELD DAYS OF my life, magical from the sheer beauty of the place. Volodia, Roma, Anton, Seth, and I piled into a small Bell helicopter, the kind where the cockpit is like a round goldfish bowl and you can see everywhere, and we flew up to the high plateau north of Norilsk, near 69.882°N, 88.76°E. The views of the sweeping intense green tundra and the huge shining braided rivers with fields and bars of snow that last through the summer here thrilled me utterly. I was glued to the window. We also saw, of course, the vast pollution of the smelters, and the endless pipelines and mine tailings.

Suddenly, we reached the plateau. This plateau might have been a fantasy movie set, a kind of imaginary haven. The land was covered with a uniform green lawn of short wild grass and moss, gently undulating off into the invisible distance. No clumps of taller weeds, or even a bush, marred its perfection. No mountains or forests rose in the distance. We had sweeping greenness forever, capped with a perfectly cloudless blue sky. We spent the day hiking the edge of the plateau and collecting the rocks that outcropped there, overlooking the silver shining river far below, edged with snowfields. These samples were mainly for Roma and Volodia, not useful for our specific investigations. So for me the day was a beautiful, relaxing hike. At the end of the day, we walked all the way down the side of the plateau and through the willow thickets to a gravel bar at the side of the river where the helicopter waited.

And then it was over, and it was time to go home. That was our last field season in Russia.

WE ALL DUG INTO ANALYSIS OF SAMPLES, INTERPRETATION OF DATA, and comparisons with each other's work. Some of us really did work across disciplines to produce data and ideas that could not have existed had we worked alone. The author lists of some of our publications show that work: Russians and Americans and Norwegians writing together; climate experts and petrologists writing together. In other cases, personalities clashed and people didn't want to collaborate. And then, in some cases, people took their piece of the pie and went home and did what they would have done anyway. This was true of the geochronologists, who did not invite others to collaborate and coauthor. Their work, however, was the key to all the rest, the key to proving causality and not coincidence.

Henrik Svensen and Sverre Planke, my original Siberian field companions from the University of Oslo, along with Ingrid Aarnes and Kristen Fristad, also from Oslo, and Alexander Polozov from Moscow, with their collaborator Nick Arndt from the University of Grenoble in France, worked together for years. They discovered and proved that the magmas heated the rocks they passed through on the way to eruption, and those heated rocks sweated out vast quantities of climate-changing gases.

Then, we needed to be able to show that some of those gases were also carried by the magmas themselves, and delivered high into the atmosphere where they would change climate and not just rain out quickly back to the surface. Ben Black, as part of his Ph.D. thesis at MIT, and I, with collaborators Ingrid Ukstins Peate and Michael Rowe, were able to measure volatile elements in melt inclusions from samples across the province, showing the magmas themselves also carried climate-changing gases. Along with carbon and sulfur, which we had expected, we found that the magmas carried surprisingly high levels of fluorine and chlorine. These levels could not have come from melting in the Earth's mantle; they had to come (as we subsequently proved) from the sedimentary rocks and hydrocarbon reservoirs that the magmas traveled through and rested in on their way to the surface.

Together with Ben Weiss at MIT and additional team members, Ben Black was also able to use magnetic measurements to show that some of the volcaniclastics had been erupted at 600 degrees Celsius or hotter. They were not the result of magmas interacting with lakes, or the result of later weathering and rearrangement of the grains. They were erupted hot and fresh and would have been the result of towering eruption columns ten or twelve miles high, equivalent to explosive eruptions today but not previously suspected to have existed in the flood basalts. And by carefully measuring the isotopes (different weights) of sulfur in the magmas, we were able to show that the volatiles came from the crustal rocks, not from the mantle where the magma first melted.

Thus, we had proved that these normally quiet flood basalt lavas carried and released a world of trouble into the end-Permian atmosphere. And that trouble came in the form of carbon dioxide, sulfur, and halocarbons, the same family of chemicals now banned by international treaties because they were destroying our ozone layer. Next, we needed to show that the gases that the magmas and lavas released were enough to wreak havoc on the life on Earth.

Anya Fetisova, Roma Veselovskiy, Volodia Pavlov, and their larger team led studies using sensitive magnetic measurements of the flood basalts from all over Siberia to show the episodic tempo of eruptions. The Earth's magnetic field moves a measurable amount every year, and so by comparing the magnetic directions recorded in their samples, they could tell how many eruptions occurred close together, and when gaps in eruption came. The majority of the flood basalts were erupted in about a half million years, but during that half million years only ten thousand or so years held the majority of the eruptions. The timing was needed to understand the atmospheric effects of gas release, which was the next step.

Ben Black and I teamed with some of the world's experts on modeling climate, Jeff Kiehl, Jean-Francois Lamarque, and Christine Shields at the National Center for Atmospheric Research. They had already created a computer model designed for the end-Permian. We put into the model the amounts and locations of the gases the magmas produced, and ran the model to see what happened. The halocarbons released by the Siberian flood basalts would have destroyed as much as 70 percent of the Earth's ozone, worldwide. And the sulfur compounds would have made rain in the Northern Hemisphere as acidic as lemon juice.

Jon Payne at Stanford and his colleagues studied new outcrops in China and elsewhere around the world that carry evidence of the oceanic extinctions. Their work demonstrated the speed of the extinctions and the episodic nature of recovery, with aquatic creatures beginning to thrive again and again, only to be killed off or squeezed into miniature versions of previous species, as the climate oscillated for millions of years before stabilizing again. Measurements of the organisms' shells also supported Ben Black's work on acid rain: the oceans were acidified again and again.

In the rock record at the end-Permian the global budget of carbon changes precipitously. Like most other atoms, carbon comes in several weights, called isotopes. The ratios of these isotopes in the atmosphere, and therefore in the carbonate molecules that mollusks use to build their shells, and the carbon-rich sediments at the bottom of the ocean, vary over time depending upon where the carbon is coming from. Carbon from the mantle, released by normal volcanic processes, is richer in heavy isotopes. Carbon that has been plucked from the air and soil by plants is lighter, and so coal and petroleum are lighter. The whole Earth at the end-Permian was suddenly overwhelmed by light carbon, causing an "excursion" to light isotopes in the global carbon record. Lee Kump at Penn State and his student Ying Cui were able to model the amount of carbon needed to create this excursion, and to define its possible sources.

About five years after our last Siberian field trip, as we analyzed those volcaniclastics, I, Ben Black, Roma Veselovskiy, and Steve Grasby, Omid Ardakanian, and Fariborz Goodarzi from the Canadian Geological Survey found within those rocks from the Angara and the Nizhnyaya Tunguska microscopic evidence for high-temperature burning of coal. For all the field seasons we were in Siberia we found no compelling evidence for significant burning of coal, when we looked at places where magma and coal had lain together on the surface. But hidden among the tiny mineral grains of the vast volcaniclastic cliffs in the south were hundreds of hard black spheres that showed magmas interacted with coals in the subsurface and combusted them within the eruption columns and plumes, sending their carbon around the world and throughout the atmosphere.

Finally, now that we knew the eruptions produced enough climate-changing gases to cause extinctions, we had to demonstrate that the eruption and end-Permian extinction were truly related in time. Seth Burgess and Sam Bowring set out to improve the date measurements and demonstrate that the volcanic eruptions began, and then the extinction happened. If the extinction happened first—well, we are pretty sure that extinctions don't cause volcanoes, so we would have been left with an astounding coincidence. Though this simple sentence cannot do justice to the many years of effort it took, Seth and

Sam first developed new laboratory techniques to increase the accuracy and precision of U-Pb dating beyond any lab in the world at that time. They showed first that the extinction began at 251.95 million years ago (Ma), and its duration was not more than 60,000 years; it could have been any number of years less, the extinction could have taken as little as 100 years, but even their extremely accurate and precise techniques could not constrain it further. They then dated the extrusive rocks (those that erupted onto the surface, rather than cooling belowground) beginning by 252.4 Ma, and continuing through at least 251.4 Ma, though the majority were complete by 251.9 Ma, just at the time of the extinction. Thus, the eruptions completed a majority of their volume before the extinction occurred.

We now know, through the interlocking work of all these people and more, that the Siberian flood basalts did cause the end-Permian extinction, through changing atmospheric chemistry with the same chemicals that humankind is releasing into the atmosphere today. Along the way to this new knowledge we trained many young scientists, we created strong and lasting ties among scientists in eight different countries and across many disciplines, and we helped produce a film that was shown on the Smithsonian Channel, along with many pieces of popular writing in several different languages. We published more than sixty peer-reviewed scientific articles and to the utmost degree possible with current scientific technique, we proved that the Siberian flood basalts produced climate-changing gases frighteningly similar to what humankind is producing today, and that in the end-Permian, these gases caused the death of the vast majority of multicellular life on Earth.

Chapter 7

THE KINDS OF THINGS A PERSON CAN WANT

The slate terrace at the back of my Uncle K and Aunt Eleanor's Philadelphia house was ringed by boxwood bushes and looked over a shallow reflecting pool framed by bowing dogwood trees. Uncle K had been called so after his middle initial since childhood, because his first name, Frank, was the same as his friend Frank Borden, Eleanor's brother. A few of Eleanor's orchids, seeming to me the pinnacle of interior design and plant care, stood on the low table between us. Uncle K poured from the ritual bottle of Veuve Clicquot champagne, Eleanor's favorite, and now mine as well. Alert, pleasant, interested, and adept at conversation, they, I felt at once, loved having me there, and also that their level of manners was beyond my capacity. And yet I aspired. We were chatting before dinner about the job I had just accepted at the Carnegie Institution for Science.

K had held big jobs. He had been the CEO and chairman of the board of Penn Mutual insurance, where his astonishing office suite, complete with sitting room and shower and bath, looked straight down on the Liberty Bell in Philadelphia. He was a consummate professional in all I had witnessed: his calm expression and level, firm voice spoke of his naval background and grace under pressure.

In the eighteenth century, Philip Stanhope, the Fourth Earl of Chesterfield, once laid out in a letter to his son his recipe for a successful diplomat and statesman: Volto sciolto, pensieri stretti. *An open relaxed face, and all thoughts kept to oneself.* Uncle K may have achieved something better: clarity throughout, with no need for artificial concealment. I yearned for his composure.

Thinking of some of the hard decisions I'd made already, I knew I had more ahead of me as the leader of a research campus. I also knew that some would involve an unhappy and possibly combative person across the desk from me. Even thinking about that imaginary angry person made me nervous. I asked K how he did it. How did he make the hard decisions, how did he face the people he needed to tell bad news?

K said he'd never been nervous in those moments. Once he knew he was making the right decision, knew it in his gut, then there was no emotional distress. Amazing, I thought. Imagine making the kinds of decisions he'd had to make, decisions that affected thousands of people. But I could immediately imagine the feeling of knowing in your gut a decision was right. I'd had that sense of clarity—and it really felt like it was in my gut—for decisions like stopping riding at that stable after we sold my horse, or knowing I should accept Ben Black into graduate school. I still could not imagine, at that moment, being able to reach clarity on the plethora of decisions that would come at me as a director, decisions that would come whether I was ready or not, whether I was feeling calm and centered or feeling exhausted and jangly.

At Carnegie I would be leading the storied and respected Department of Terrestrial Magnetism, or DTM, one of the institution's six research units, all reporting to a single president. The Carnegie Institution for Science is one of the twenty-odd organizations named for its founding philanthropist, Andrew Carnegie. Established in 1902, the institution is endowed and designed to provide scientists with the support needed for them to conduct research without being beholden to or influenced by their funding agency. President Theodore Roosevelt served on the founding board of directors for the Carnegie Institution. Edwin Hubble, who discovered that the universe is expanding, and his scientific descendant, Wendy Freedman, who has made the most exact measurements of that expansion, worked at Carnegie. The organization was the inspiration for the National Science Foundation, and it achieved and continues to achieve many major breakthroughs in human understanding.

I became the seventh director of this department since its founding, and the first woman. I was responsible for the support of fifteen permanent research staff, people in positions similar to professorships but with no teaching component, since Carnegie is not a degree-granting institution. At its monkish campus in northwest Washington, D.C., the scientists are free to think and work largely uninterrupted, while supported by a significant staff. I was expecting to spend time in leadership, and also to continue my own research program.

I knew there would be great pleasure and collegiality there. I also anticipated that, as director, I would naturally need to help personnel overcome deficiencies in their work, settle disputes, and make budget, staffing, and policy decisions that would undoubtedly displease some. I might even need to fire people. I thought a lot about Uncle K's advice on knowing my values.

The communications, team building, writing, selling, marketing, forecasting, and budgeting that I learned in business were suddenly invaluable for running a lab group at MIT, and they were critical in winning and then executing my eventual leadership positions. Those years in business had given me a basis for the technical side of management, but the next decade would greatly stretch my abilities on the human side. Culture, that is, the ways we all treat each other, would be key in determining the success of the teams I'd be part of.

ONCE I REACHED A CERTAIN LEVEL OF SENIORITY, I'D OFTEN BE ASKED by students for advice. This is an experience, I imagine, that is shared by underrepresented people in all sorts of fields. When I was a postdoctoral researcher in my early thirties, one graduate student came to discuss a problem she was having with her advisor, who seemed to have taken her data for use in a paper first-authored by someone else (the first author, in Earth and planetary sciences, is the person who leads an effort). She told me that she believed he regularly searched her desk and looked at her computer files. She was enraged, but also afraid. If she complained, and her advisor retaliated, it could end her career.

She and I reviewed the many possible responses she had considered: talk to her advisor about it; talk to her department chair; talk to the other faculty, with other chairs, with the dean; work to better document the problems; set booby traps in her desk; leave for another university; allow it to continue uncontested. What did I know about solutions to problems like this? I'd faced some issues in my own life that felt a bit similar, and I could share what I had done and what had happened. But I didn't know with any certainty what the right answer was; I only knew that some of the ideas felt good, and others felt bad to me.

I thought of a way to describe that good or bad feeling. We asked ourselves: Which of these actions would make you proud of yourself, when you look back at this problem ten years from now? I've used that guideline many times since. If you are proud, looking back after ten years, then that is the high road. Always take the high road. Don't chicken out, and don't go for the gratuitous instant emotional reward, which is almost always ugly and petty when looked at with the clear eyes of tomorrow. Take the high road.

The irony of these thoughts struck me: I had been asked several times, upon my departure to the business world from academia, and upon my return, how I could have stood working in a community—the business community—that was so devoid of ethics and values. My own experience is that culture supports or destroys ethics, and that cultures of all kinds exist in all human sectors. There are deeply ethical, caring, principled cultures in business, and disrespectful, ego-driven, bullying cultures in academia, and there are the reverse as well. How much, I wondered, would the culture at the Carnegie Institution for Science change me, and how much could I change it?

THE FORMALITY AND IMPOSING HISTORY OF THE PLACE STRUCK ME as I walked up the tall granite front steps of the main building of the Carnegie campus for the Department of Terrestrial Magnetism, and their sister organization, the Geophysical Laboratory. This beautiful ten-acre campus in one of America's bastions of privilege, Chevy Chase, Washington, D.C., is a peaceful, sequestered preserve of thinking and research. Within the Department of Terrestrial Magnetism itself Vera Rubin, who won the National Medal of Science for confirming the existence of dark matter, was still at work when I arrived. She retired while I served and thus I was her final director, and I was also her friend. The department had been founded in 1902 to make the first comprehensive maps of the Earth's magnetic field. The Abelson Building, whose stairs I was climbing that day, was built in 1914. To reach my corner office with its views of both flowering cherries and dogwoods and the metal dome of the Atomic Physics Observatory, which once housed a three-million-volt Van de Graaff generator that created one of the first demonstrations of uranium fission in the United States, I could use a private door, or enter the way visitors did, through the anteroom and office of Jan Dunlap, experienced assistant of directors before me. Jan welcomed me with great warmth and also with some fairly strong opinions of who a director should be and how they should act.

One of our early disagreements was on forms of address. I prefer everyone to greet and refer to each other by first name. Removing the honorifics makes a more level playing field for everyone to contribute and be judged on the merit of their ideas. Though there is good reason to keep the "Dr." or "Professor" in front of the names of people who traditionally have been excluded from those roles (women, people of color, and more), still, then presumably equally able and participative non-degreed team members are immediately relegated to a lower rank. So, whenever possible, I like everyone to use their first names. Using first names was a departure from practice in the department and caused some waves.

Then, there was the simple fact of my sex. There had never been a female director. My predecessor, Sean Solomon, always dressed formally in a suit and tie, and brought formidable gravitas to the position with his scientific expertise, his demeanor, and his air of authority. What would be my equivalent style of dress, of conversation, of decision-making?

I looked at the forms Terry Stahl, our business director, laid out, all on Carnegie DTM letterhead and organized impeccably. I asked him for a cash flow statement that would show when and from where our income came, and where it went, each month. "Oh, we don't do it like that," he replied. I was to hear that phrase a lot. Tradition carried the day at Carnegie, except when it was important to forge a new path, and then, I found, I had the confidence of my convictions to lead that way. Over time, Terry and I created an excellent working relationship with, I think, mutual respect. I think he realized that I was not questioning his expertise, but instead I was questioning the half century and more of unchanged management practices.

I pushed forward slow step by slow step, both finding my way with new initiatives and working on long-standing problems like lack of overtime pay for the building engineers, and removing a facilities manager who sexually harassed his cleaning staff. I was getting over my own internal conflicts of whether or not I could lead, and every day I felt clearer about leading as my own person and not in the mold of someone else. Reactions and actions, like the use of the honorific "Dr.", made it clearer all the time, like a box of sand shaken until the buried stones emerge on the surface, that every adult is conflicted at least subconsciously about the roles a woman could have.

My mind went back to an even earlier time with Uncle K. I was in my twenties, and had launched that little company creating business plans for high-tech companies. Though I still tended to dwell upon the failures, I knew I had a lot of successes, too. I was hired by Johns Hopkins School of Medicine to write business plans for two of their very earliest medical device spin-off companies. Some of my clients respected my skills. One of my clients, after I had helped them on a couple of smaller projects, offered me an opportunity to help run a small aircraft company. This made my heart race. Secretly, I really wanted this. I wanted the step up in responsibility and in respect, and I wanted the challenge, and I wanted to walk in the footsteps of my illustrious grandfather John Tarbox, who patented the first mechanical autopilot for a plane, and consulted for Curtiss Aeroplane and Motor Company. I went to my Uncle K for advice.

We were, again, out on that stone terrace, the site of so many momentous conversations. We each had a bright cool drink on the wrought iron table between us. I mentioned the offer I had, a bit off-handedly; I couldn't really own it even in a story. He thought that aircraft company leadership position was such an outsize opportunity for such a young woman that he literally laughed until he cried. The idea was so ludicrous to him that he didn't even talk about it; I think that he thought I agreed and had brought it up as a throwaway comment, something impossible. But I never asked him why he laughed. I was both hurt that he would dismiss the opportunity out of hand, but even more worried that he was right. I put it out of my mind—I did not pursue the opportunity any further. And so I struggled with knowing what I wanted, what I thought I could do, and what the world would allow.

A few years after Uncle K laughed until he cried, I left consulting and went back for my Ph.D. There's a myth that the people with high academic research achievement got there through an inherent disciplinary genius and a drive that came from childhood. Well, inasmuch as I can be considered to be highly achieving, you can see that I was not blessed with an inherent disciplinary genius or a drive from childhood.

My colleagues and the head of our department were not unaware, however, of my departure from academia during my twenties. I was advised to minimize those years on my curricula vitae so long as I was still junior faculty, that is, not yet tenured. I was expected to give every appearance of never having veered from the academic path. I was told not to show the reference book series I had written; these books, aimed primarily at high school and college students, were not considered helpful for getting tenure. Academia wanted me to look laser straight and undeviatingly rigorous in my work: outreach, public writing, and primary and secondary school work was considered noble but a dangerous sign of possible weakness in dedication to the necessary quest for the most important and difficult scientific discoveries. Further, the organizational and financial skills I had gained were not valued at the professorial level.

Finally, in my forties, all those skills had found their day. I needed those organizational and financial skills, and I needed all the scientific expertise I had gained. And at last I understood to my intense relief that I did not have to become some different kind of "leaderly" person—someone perhaps more stereotypically dominant in dress, attitude, and behavior—when I became, in fact, a leader of an organization. I could remain my truest self, and in so doing, also speed my path to clarity and best decisions and practices. My truest self still sometimes includes sparkly nail polish and black leather boots, and often includes laughing.

AS I WALKED UP THOSE GRANITE STEPS TO THE ABELSON BUILDING that day in 2011, so had I ascended to a new level of leadership in science. I thought, at the time, that I had taken the position because of its challenge and its opportunity to make change, a chance I would not have had at MIT for at least another decade. I had taken it to experience a new level of freedom of decision and a new level of influence. All those things brought me happiness.

The much bigger salary was also a lovely present. I had never thought very much about salary, outside of a wish for equity with colleagues. When I'd been a full-time math lecturer at St. Mary's College of Maryland, I had been told by a colleague the salary of another lecturer over in physics. He was making almost twice what I was. I felt my blood rise up and flush my face with fury and with embarrassment. Shown up to be weak and unworthy again. And, mad as hell. I went to my department chair and asked him about this inequity. He seemed a bit shaken, as if the unfairness of it was suddenly obvious to him but he had never noticed it before, and was caught off guard. But I am just guessing at his thoughts.

"Well," he responded after a moment's pause, "he's completed a couple of years toward his Ph.D."

And I said, "But I have completed my master's, and he has not."

"Well . . ."

And that's all there was. I did not see, at that time, another path, and I simply put that inequity into a pile with the others. I'm a bit concerned now that as that pile has grown over the years, the path to my anger has shortened and heated up. So getting a big raise from my MIT salary (which, incidentally, was set to be identical for all entering new professors in a given year, a key step for equity) was welcome indeed but not my driving purpose.

Standing in our new kitchen in Washington, a kitchen I loved in a house I loved, not least because I myself had bought it and I myself had furnished it—James had not been able to come down for house hunting because of the constraints of teaching, and so I had shared decisions via videophone—I was talking with my brother Jim about the new job. Jim brought up his sense that he had risen to the top of where he was going in his field. I felt a little chill, a little ripple of foreboding. To me, he had always been exceptionally successful, and also confident, almost preternaturally self-assured, in his decisions. He'd published over thirty books. He was world-famous in his field, art history, far more so than I was in mine. He had a named chair at the Art Institute of Chicago, and had turned down many offers to move elsewhere, a kind of metric of success in academia.

So much of my learning was thanks to Jim. He'd given me books in junior high, and high school, and college, and through my twenties, that changed my life. Because of him I read Elias Canetti's *Crowds and Power*, and discovered Miłosz's poetry, and his great *The Captive Mind*, and learned about voice from Louise Glück and Kathy Acker, and broadened my views with the classic Chinese novels *The Journey* to the West and The Story of the Stone. He gave me Robert Musil's The Man Without Qualities and I still think of that paradigm when analyzing lives. As I started in science, he prepared me with Thomas Kuhn and Karl Popper. He taught me about music as we played flute and piano duets, every day, for years; I'd come home from school and there would be a new score on my stand, ready for sight-reading. He borrowed the score from the Cornell music library and we listened and followed until I knew Bizet's Carmen entirely, in preparation for attending my first opera. He taught me about art, and museums, and criticism. He challenged me to write a paper comparing works by Max Ernst, Norman Rockwell, and Jean Dubuffet, who remains one of my favorite artists for his early work. Thanks to our conversations I eventually created the NASA Psyche Inspired art intern program, to bridge and interpret space exploration, science, and engineering through the arts. The richness of the fountain of his giving opened the world to me.

The crux of the matter that day, it seemed, was that my new salary was high by academic standards and he'd noticed it. Suddenly I felt I was floundering a bit. We had been just chatting, standing on the bank of the river, so to speak, and suddenly I was struggling in the cold water looking for a ladder. Should I apologize? I think I said something vague about the inequity of salaries between the humanities and sciences. But that day I knew a boundary had been crossed, and it had been crossed by me, from the place I was supposed to be in to a place I wasn't supposed to aspire to. I'd crossed that boundary before in the eyes of some who thought women were not scientific leaders, but it had never occurred to me that such a boundary existed between my brother and myself.

That feeling of being in the cold water all of a sudden, the plunge into the wrong medium and the thrashing to get out, stayed with me. I had never contemplated a world in which my own family would not be purely happy for any successes I managed. All the years my brother was so much more successful than I was, while I contended with depression and got my act together and became a better mother and went back to graduate school, I had only unalloyed happiness for his career advancement, and everyone in the family was loving and supportive of me while I struggled. I was as blind as, perhaps, only a little sister can be.

That day, I learned I was, in fact, a competitor. I had mistaken my brother for a parent. And I wondered, if you love someone, aren't you necessarily, by definition, also happy for them when things go well for them? I think so. And so I wondered both whether love was too fragile, and whether I should have kept my work from my family, or at least curated my stories much more intensively. I realized that I should have.

As a little sister, and as a woman, I was not supposed to succeed and exceed in the ways I did. One of my aunts regularly comments, no matter how little I manage to tell her about my work, "I think you have succeeded beyond what your mother would have thought seemly." Whatever the measure of success I'd reached, it made my brother and my aunts step back, and even made my son a bit unhappy, at a time when he was building his work but not yet at the success stage. When I learned this, I felt my heart had broken, and my sense of strength and joy at striving dissolved into a kind of grayness. I wonder if other women and members of other undervalued groups have been through similar experiences; I wonder if I could have learned faster and better handled my relationships.

What was I trying to attain, anyway? Gaining leadership jobs, this one and additional ones to follow, felt like a path to freedom for myself and, I hoped, for the teams I worked with. They gave me a kind of stature to talk about what matters: opportunity, respect, culture, achievement.

Achievement. In America, we are free to be in pursuit of happiness. We also often pursue, and idolize, power. Happiness and power. Happiness, some assume, comes at the moment of attaining success, and success is often defined by power, money, or fame. But happiness gained by success is an exceptionally fleeting emotion, attained only for the moment a goal is reached, then slipping away again. For this reason, we may be free to pursue happiness, but if we mean to achieve it via success, I think we have a dead end.

In academic science, what is success? For the first decade and a half of one's academic career, the most basic measure is obtaining tenure, and thus being assured of that academic position in perpetuity. (Unless you transgress laws or policies egregiously, and often, not even then would you lose your position. See the #MeToo movement and its varied levels of success with tenured faculty.) Some people get tenure and then experience a feeling of loss of meaning. Why struggle onward with new scientific discoveries, when only a few people in the world really care? What do I have to work toward now?

As I have become more senior in my field, I've become aware of a couple of friends who were secretly wishing and waiting to become a member of the National Academy of Sciences, one of the highest accolades we have. Each of them was consumed with joy when their moment came—and I am sure there are many I know who have not shown me their feelings because they have not been elected—and then their joy subsided over the following year into a more usual cynicism and disaffection with the world. They were experiencing what Peter Matthiessen, in *The Snow Leopard*, described as "the desolation of success." Success and its presumed partner, happiness, are ever-receding.

It is better to set the goal that can never be accomplished. Then you never reach the existential void of success. And that's why winning the Psyche mission felt so good. It was the beginning, not the end.

For a while I was puzzled by my internal state of not wanting—in fact, having a visceral reaction against—being told I am exemplary or successful, and at the same time, hating being underestimated or condescended or lectured to. Am I just being overly fussy and self-absorbed? I wondered. Well, maybe. But then I realized the underlying motivation: the fundamental unfairness of a hierarchical organization is that people at the top get a disproportionate amount of credit, and the people several layers down get too little credit. I wanted us to create organizations where each person gets the credit and the responsibility that they have earned. Now that I was in leadership, I was getting disproportionate credit.

Hierarchical organizations are designed in part, ironically, to ensure that credit, responsibility, and indeed, any kind of attention, comes to each person according to how present and critical they are in the eyes of their coworkers; that is, it comes in proportion to the amount of focused attention each person gets. Hierarchical organizations are built in some cases for the specific purpose of having a leader for each relatively small group of people, a group of a size that the leader can pay attention to and learn about each person in their group (some organizational people say that number is six). In this model, the leader's mandated job is to pay attention to each person in their group. And yet, as if each layer of hierarchy is an insulating blanket instead of an attentive sub-team, people lower in the organization get less credit and are less heard.

So what was I after, really, if not success and fame? Happiness, and power, both, I am sure. But having not expected any of the usual metrics of success, coming so late to an academic career, I had not pinned my happiness upon them. After significant reflection, James and I came up with a list of three things that lead to our happiness: who we wake up with in the morning (each other, and in our heart, our family and friends), where we are (the landscape, the city), and who we work with during the day (the team, the people we spend so much time with). Those, and a perfect slice of grapefruit, or a lot of morning glory blossoms early in the day, or identifying a species of bird new to me. Now, I have moments when I can feel I have some power, but mainly, I experience that as responsibility. I have more responsibility to more people.

Think about what we can want. We can want happiness and power, but we could also purpose our lives toward amazing things like equality, like being listened to, like a chance to contribute, like justice, transparency in leadership, collaboration, and community. How I wish to live in a world where people don't condescend to each other, and I've become more and more intolerant of condescension toward me. So perhaps I have, with leadership, a chance to move the world a bit in that direction.

When it came time to leave Carnegie years later, James and I made a table on a piece of paper on our refrigerator. Down the left side we listed the key characteristics we wanted in wherever we next moved, and along the top, we listed the job options I was considering. We ranked them in each of our key characteristics. James, his international mathematics education consultancy having ramped up, could live anywhere and still do his work. Our key characteristics included aspects of the town we'd be in, including how close it was to Turner and Liz or how easy to travel from, and it included the culture and nature of the team I'd be joining.

No matter the family tensions and the advice I get to work less, I really love my work. I don't want it to be less of my life. And so, I began wondering, along the way, whether a motivated team with a tremendous vision might take the place of a family, at least a tiny bit. I thought that these people with whom I spend the greatest part of my waking hours could provide some of the loyalty, purpose, and support that a family ideally gives. I'd already seen it happen, and experienced the joy that kind of team experience can bring, and I'd seen the awful disruption when a team failed. Maybe, I thought, the impact of a work team is so profound that it's incumbent upon us to make it a culture as positive for each individual as we can. Maybe that is something we can want: the ability to support a team, and be supported by a team.

Chapter 8

EXPANDING COURAGE

I sat in the chancellor's waiting room, decorated with Turkish carpets and pale wood bookcases filled with awards, on the top floor of a tall academic affairs building. Snow fell past the window on its journey down to the street far below. My hearing was acute, my muscles ready, all the symptoms of adrenaline that came with the inevitability, the necessity, of another frank conversation verging on confrontation. Of course I sat calmly, not fidgeting, embodying the courage I was constantly cultivating in those days. The executive assistant was typing on her keyboard, having kindly offered coffee or water, and I was fully in the embrace of all the dignity, tradition, and power of the organization I was a part of, and now, pushing toward change.

The chancellor welcomed me in with his usual friendly smile and his boardroom manners. He asked after my family, commented on some recent events in the news, and ushered me to sit on the couch. He took a chair across a low coffee table in the splendidly furnished office with its Oriental rugs and art. This familiar guest couch verged on enacting a power play of its own: it was low and soft, and so I always sat a little awkwardly, and also below the chancellor's eye level. I vowed to sit on one of the firm and tall end chairs the next time.

"We need to reconsider this conversation about Chris," I began. "The complaints as you have them recorded are far more mild than what has been reported to me."

"Yes, you know, I've been thinking about this a lot," he replied. He went on to tell me that he had discussed the ideas with friends, and, based on what they understood had transpired, they all felt this was like social hugging: some people are comfortable, and others just don't like it. The actions the chancellor was comparing to social hugging were allegations of fondling and kissing someone against her will.

"Respectfully, social hugging is not a metaphor for this behavior," I replied, realizing with a sinking sadness that he was determined about his position. "Fondling and kissing a direct report is sexual assault." I could see on his face the distaste he had for my bringing up that term. I'd crossed the line of polite conversation, and begun to push him into a legal corner.

"If the allegations against Chris of harassment and bullying are true, and I believe that they are, then Chris has invalidated himself as a leader," I told the chancellor. "He has to be taken out of his position of power."

I was met with a series of rebuttals, including, as I recall, the need to keep Chris because he brought in a lot of grant money, and that Chris should be forgiven because he was drunk when he did it. No, my friends, we cannot overlook behavior while someone is impaired. Later, I told that one to brilliant Sara Collina, she of the streamside party where I met Curtis so many years ago, and she quipped, "That's funny, because when I'm drunk, I embezzle." So clear, when it's put that way, that being drunk is not an excuse for illegal behavior. Sara, a friend since we were fourteen, grew up to become a brilliant, fierce advocate for equality and change, and having her put my experience into perspective over many lunches saved me that year.

THEN, IN THE DAYS BEFORE #METOO, I WAS FACING A DIFFICULT challenge. I was working in leadership at an academic organization I'll not name here. Before I had even started the job, I had been warned that Chris (not his real name) was troublesome. Though Chris was not in my part of the organization, once I arrived members of his staff started coming to me with complaints. They saw me as the closest, most accessible senior person and I felt it was my job to listen and to help if I could. Simultaneously, as a new person with no history with Chris, I was trying to forge a strong and positive working relationship with him so we could move the organization forward.

Members of Chris's staff felt he wasn't transparent with decisions and that he was playing favorites. Some people told me they had been bullied, others said they had been shut out of decisions or denied access to resources. I gave suggestions: "Have you talked with him yourself, and expressed your specific concerns, and discussed ways to remedy the problems? If that has not worked, have you talked with the HR person in your group? Have you then spoken to the chancellor of the organization, Chris's boss, about the problems?" All these paths had been taken by various people, to no avail. Others had been too intimidated to try.

Problems began to move into my arena as well. Chris didn't want to share the cost of common resources, like IT, business operations, and facilities work, in our common building. I was certain we had an agreement on a date when some of the functions my unit had been handling would move to his, and when other functions would be paid for in common between us. But when that date came, he wrote me an email stating that I must have been mistaken and misunderstood the conversation, since his unit was much too overburdened to take on any more expenses. Chris also wanted to take more office space for his own unit, at the cost of mine. He could not take the rooms he wanted without my approval. We discussed many options over several weeks, and were at an impasse.

Chris was not getting what he wanted on several fronts, not even through bullying. The chancellor was unwilling to act, even after multiple meetings and discussions of various alleged wrongs, administrative and behavioral. The chancellor wrote to me and said he didn't have any ideas on how to fix this. And so I was stuck between both of them and the faculty and staff.

Staff complained to me that they felt bullied by Chris, and women in his unit were reporting to me that they had been disrespectfully treated. But then one evening at an on-site party, Chris had gotten drunk, had inappropriately touched and pressed against a woman while dancing, and later, it was reported to me, had forcibly kissed and fondled a woman on the staff, in front of other people.

"Wait a moment," I said on the phone when the woman first told me, "could you go back to that part about the kissing and fondling?" If the allegation was true, that was sexual assault.

I called the chancellor and reported what I understood had happened. In my world, someone who was accused of having done these things was escorted off campus and put on leave until proceedings could be completed. In the weird world I found myself living in, Chris and the female staffer both stayed in their regular positions.

In the following days I had staff coming into my office offering to picket the administrative offices, to write editorials to major newspapers, to go to our professional organizations. Some of these people told me they had witnessed the events at the party. As tempted as I was myself to make some big signs and march in the street, I really wanted the organization to fix this without additional harm to its mission and excellence.

My stress was at a high pitch. Every day that week going into the office I had a sense of dread: *What is going to happen today?* And every day I had to exercise all the advice I had so often given to others myself. I had to think about taking the high road. I had to keep my ethical core clear. I had to stay calm in meetings, and listen and listen before speaking. Of course, I did not always live up to my own standards, and I felt the stress eroding my well-being, eroding my sleep, and health, and ability to get my own job done.

After ten days, nothing had happened. The woman who told me Chris had assaulted her still reported to him, and she asked to talk with me. She asked me what she should do. She had been using sick days to avoid seeing him or working with him. The next day I received a letter from the organization's lawyer telling me to stop checking in with this woman staffer, and not to discuss the events with anyone, including her.

I emailed the chancellor again, asking for action, and received that meeting in which he told me Chris's alleged behavior was like social hugging. Every night I'd walk up the tall brick steps in front of our fancy brick house, the one I'd bought thinking we were going to live here for decades, and kind and infinitely understanding James would hand me a glass of red wine and ask me, "What happened today?" There was always a new story. Sometimes it was about Chris or the chancellor, and sometimes it had to do with other problems I'd inherited and was seeking to sort out. We'd sit in the living room on our sofas and James would listen.

As much as James's and Turner's support and advice helped me sort through the confusion and stress of that time, I also increasingly felt I needed advice from a more experienced source. Sara, herself a lawyer, knew just the practice to help me parse my options. Soon I had set up an appointment to get some professional advice from a lawyer deeply experienced in human and workplace rights. This lawyer explained to me that there was a law where we lived that said that the senior cognizant officer of an organization who had taken no action to fix workplace harassment and other similar issues could be personally liable for the wrongs. In other words, if my organization did not make things right, and one of the employees who had complained to me about Chris sued, they could also sue me personally and go after my retirement, and my house.

We sorted through the risk. First, most of the staff knew I was working every day to try to make this right. I had documented all I was doing. And we knew the endgame was to go to the board of directors; it was becoming increasingly obvious that I was going to have to take that final nuclear step, because of ethics imperatives and legal imperatives as well. Life shrinks or expands in proportion to one's courage, said Anaïs Nin, and I felt that I was helping my life expand.

In response to my insistence at my meeting with the chancellor, the organization brought in an external lawyer to interview everyone and write a report. Over a period of weeks, this external lawyer interviewed us all. I knew the stories; at this point the ones that predated that party had been told to me over a period of a couple of years. But, as I saw it, the report whitewashed all of them. Chris was allowed to keep his position. And he was officially exonerated by the organization.

I now knew of two men who were bullied so badly by Chris that they had gone into therapy to contend with the stress. I continued to hear reports of harassment and bullying. Yet, I had been told to take my complaints no further. I now anticipated that I might be fired and I contended with a mix of dread and resignation that my career could be ended by this.

But despite that fear, I could not tolerate the situation. I don't think this intolerance is a virtue, just some kind of fierce compulsion. My gut was clear about this. I could not walk away. James and I talked it all through. I could stay on and comply, which meant no change and no justice for the people who had been wronged. I couldn't imagine tolerating that. Or, I could just leave. That seemed possible, but also, deeply dishonorable. Or, I could stay and keep trying to have some action occur, even though I had been ordered to stop talking about the problems. I decided I had to take that step. As United States senator Elizabeth Warren notably said: "You don't get what you don't fight for."

I took the issues to the board of directors. I could not tell, when I spoke on the phone to the group of three directors I most trusted, whether they believed me, or even felt this was a problem of strong interest to them. I was yearning for some personal support from them, but they were properly noncommittal. Having support from so much of the staff was fantastic, but having virtually no spoken support from the administration was deeply discouraging and even enraging.

I wanted their empathy and support like a dry plant wants water. I wanted them to take the task at hand seriously, and also to mirror back to me what I was risking and what I was feeling. I wanted them to acknowledge that I was summoning courage for this work, work I viewed then and still view as necessary and also agonizingly stressful. None of that happened. I received their valuable time and attention to the problem, for which I am grateful indeed, but I was expected to do my job correctly and without sympathy. Fair enough, but then, as now, I still yearn for the recognition of my personal experience. Then, as now, my dedication and connection to my team feels a little bit like my dedication and connection to my family. I expect that we value each other and that we support each other. The notion that supportive behavior is only because of a paycheck is confusing to me; we treat each other kindly and well because we have a common purpose, and because that is just how people should treat each other.

When I was a kid, I thought there was a universal standard of justice. If someone was clearly in the wrong, they would be stopped. What I have learned is that there are a million reasons why people are not stopped. The more powerful the person and the more people around them benefit, the less chance they will be stopped. Taking action against a wrongdoer is painful, difficult, unrewarding work. As Martin Luther King said, many of us prefer order over justice.

Cases of harassment and bullying are almost never clear-cut, and stopping the wrongdoing almost always requires the loss of things valuable to the institution, for example, the loss of a wrongdoing but powerful leader, or a funding source, or public reputation. Only a courageous and ethically clear leader can do that, someone who puts that sense of justice above other virtues, like money, success, and appearance, and we did not have that person.

But just as there are so many reasons why fallible people fail to take action toward justice, there are so many good reasons that they should find their courage. The three main ones, as I told the chancellor, are: One, ethics: allowing harassment to go unpunished is unethical. Two, productivity: allowing harassment to go unpunished harms the culture and lowers the value and quantity of work being done. And three, legal: allowing harassment to go unpunished places your organization at risk for damaging legal action. For reasons I still only partly understand, none of these arguments gained any traction. When I discussed these issues with my father-in-law Buff, deeply experienced in institutional leadership, he immediately asked, "Why is the chancellor protecting Chris?" My eyes opened and my mouth shut. Why, indeed? We never discovered why.

Three more months went by, living with the status quo, with Chris in his position and many of the people who worked with him suffering, although he did have some favorites, people who benefited from his power or appreciated his scientific collaboration. By now more than half a year had passed since the incidents at the party. My stress level had not decreased. Several of us on staff joked we had "Chris Syndrome": we were looking forward too much to a glass of wine at night, and our exercise regularity had dropped off, and we'd each gained some weight.

James began to feel harassed by Chris. He would appear at the coffeehouses where James often sat working at his laptop when he wasn't traveling, and mock him for having a wife working in a fancy office while he worked in a coffeehouse. James understood Chris to be insinuating things about being in the office with me while James was effectively locked out. I let the chancellor know. Again, I felt brushed off; I recall him telling me that James must have misunderstood what Chris was saying.

In the midst of the stress, I'd periodically have dinner with my father's brother, Bob Elkins. Bob and I had always been close; we'd had a special bond all my life. He was a deeply connected active listener and he was generous enough to listen to this story unfold. Over one such sumptuous dinner, which included oysters, a rich red wine, and steak, I was telling Bob the latest drama. He was astounded.

And he commented, "Being a woman in science is incredibly hard." At that moment, I was a bit taken aback, and my instinct was to disagree, and say I had been privileged and mainly supported. Both of those are true, but it's also true that as I rose in leadership the obviously gender-based biases and barriers grew. I realized in that moment that that was the topic. The topic was implicit, and explicit, bias against women in science.

When I reached this position of leadership myself I finally realized just how far we still have to go toward equity. Just how invisible the iceberg of implicit bias, harassment, and bullying is to others in leadership, and how strong their own implicit and, in some cases, explicit biases are. The experience of working with Chris and the chancellor sharpened my appreciation for the headwind women and people of color are walking against all the time.

Why are the topics of the #MeToo movement so shocking and so hard and so radical? People should not be able to assault each other, or coerce each other, or bully and belittle each other into oblivion. These are such obvious values, such obvious tenets of civilization. Who could disagree?

The topics are radical because the fighters are women and people of color. Women have been assaulted, coerced, and belittled at will in most cultures, well, forever. The role of women as second-class citizens is codified in law in many countries today. Men own the possessions, and make the decisions, and determine the freedom of women. In the West we are only a few years past that codification in some places, and in other places, not even there yet.

Changing that narrative is a Herculean, or might I say, Athenian, task. This behavior, and the expectations about who is giving and who is getting, who is deciding and who is sitting out, is baked into our brains. Not everyone even sees these issues as problems: they are invisible unless you are beset by them, and then, only visible when you have really opened your eyes and rejected the status quo. And then, who's to judge the seriousness of the problem? And whose voice will be believed?

Setting the problem up this way, in its historical context, makes it sound like a story of men versus women. But I contest that structure strongly. Men are also harassed and assaulted; men of color are underestimated; even short people are underestimated.

The true framing, the conflict we are facing, is all of us versus our implicit bias. Each one of us, regardless of gender or color, needs to constantly consciously update our automatic opinions about other people's qualities.

Let the person's actions and contributions, not your expectations, speak for them.

But here's the problem: humans tend to perform to the expectations that their group has for them. Performing exceptionally when the people around you degrade, discount, and overlook your contributions requires superhuman strength and relentless determination. So create teams, cultures, and behaviors that expect the best of everyone, and notice what each person really does.

The change we are trying to make is a change in culture, a change in expectation, that causes a change in our subconscious reactions and thus in our implicit bias. Changing culture and bias starts with new laws and new policies, but that's not at all enough—change becomes real when we see all people as equally deserving and equally promising. These ideas poured into me during that long struggle for justice in that organization, and they informed my ideas about how to create better teams.

I watched our progress, or lack of it, toward a more advanced vision of justice and equity. I watched our mistakes and failures, and formed some ideas on how to create a culture that is non-bullying and non-harassing:

- Do not divide in order to unite: When speaking about gender equity, don't have separate groups of men and women discuss it separately. Those are support groups, where common experiences can be shared. Create a group of the whole, where everyone is searching for a better common future. That is a change group.
- Make the responsibility for equity, diversity, and justice every person's responsibility. If there is one anointed director of equity, then

everyone else feels a bit relieved of the issue and can leave it to that
one person to think about.

- Think of local solutions to systemic problems, for example, having
  the department head write letters of recommendation for students
  who fear retribution from their harassing advisors.
- Listening is necessary but it is not a solution. Leaders may listen
  carefully to issues and complaints, but only their action from the
  top, and the common action of the whole team, can change the
  status quo.
- Create a granular culture that is consistent with equity. Listen to
  everyone's voice. Invite everyone to contribute. Notice the merits
  of every person, no matter their place in the hierarchy, or their sex,
  or their race. A corollary to this tenet is that a hiring rubric can
  bring in a diverse workforce, but only a culture that values people
  according to their merits, and minimizes implicit and explicit bias,
  can keep the team together.

After a final consultation with my lawyer, I went again to the
chancellor. I told him that I believed behavior like that of Chris was
intolerable in leadership. That I believed that leadership was giving
the message that women and junior personnel were unsafe, and were
also not worthy of protection. He was giving me responsibility but not
support. The chancellor had ordered further anti-harassment training, but just for the graduate students and postdoctoral fellows, and
not for the faculty or the staff or the directors. What kind of model
were we giving them, as they went out into the world into positions of
leadership of their own? One postdoc said to me, "What message am
I supposed to take from this? If I bring in enough money, I can grope
anyone I want?"

I told the chancellor that I would leave the organization unless
Chris was forced to step down. Him or me. Was it this ultimatum

that finally pushed them to action? Did Chris feel like he was leaving on his own terms, with the report exonerating him on the books? I do not know. But within a week Chris had stepped down, and in the end, he left the organization.

And so did I. A few months later I accepted a new job and moved out of town. As a friend said about his experience of being a whistle-blower in the military: "After you drop the grenade in the briefcase, it's best to step away."

SHOULD SUCCESSFUL OR IMPORTANT PEOPLE BE ALLOWED TO PERsist in a community when they are harming others? To me the answer is unequivocally no. Some of my adamancy comes, I know, from my experience of childhood abuse. But much of it comes from a perhaps utopian idea that we can create communities and teams that work in a united way to eliminate harassment, to celebrate merit and success, to minimize implicit bias, and to produce faster and better results. A mark of civilization should be that all its people can work together and be valued for what they create and produce; damaged people who use bullying and harassment have no place in this society.

Of course, we are not there yet. And now that I was in positions of leadership, I had the opportunity, the responsibility, and, alas, the regular necessity to try to serve some justice, or at least to protect those who needed it. Soon enough, I got a call from a postdoctoral scholar I knew. She had been sexually harassed by a senior scientist whom she was collaborating with; he had approached her for sex and had been physically inappropriate in ways I won't describe. After she had ended the collaboration as a result of the harassment, he had taken her data and was threatening to publish without her. She would be judged, as all junior scientists are, by the degree of her productivity and thought leadership. She needed that data and the first-authored paper it represented. She couldn't get back the months or years it had taken to produce, and her career depended upon her output.

"Lindy, I need your advice," she said. My heart sank a bit, not knowing what to expect next. "I've submitted an abstract on my research to the American Geophysical Union fall meeting, but I've also heard that he is going to be there. I don't think I can stand up at the podium and present, seeing my harasser in the audience and knowing he is against me and attempting to harm my career. I just don't think I can do it."

What an awful circumstance. The structure of academia means that the senior scholars who advise and mentor younger scholars have power over them: power of the recommendation letter, of the whispered warning to possible employers, of the devastating comment or question said in front of hundreds of colleagues at international conferences. After we finished our call, I called a leader at the AGU, which is our biggest professional organization for Earth and space scientists, with over sixty thousand members. The fall meeting is one of the major showcases of new science. More than 25,000 people come to this meeting. Standing at the podium to present your new science at the fall meeting can launch your career, or it can be a major public failure if things don't go well. I always tell our lab group that every talk at AGU is a job talk.

Luckily, I'd been the president of the planetary section for AGU, so I knew the leadership personally, and I knew them to be deeply thoughtful and purposeful. This leader, however, was not quite ready for my question: Had the harasser been found at fault by his organization? As it happened, yes, he had. But that wasn't enough. The AGU did not yet have provisions in its code to bar people from the meeting on the basis of this kind of wrong. And they wanted to decide carefully and in the right way how to handle this, which would take some time.

So the postdoc withdrew her abstract and did not attend the meeting that year.

Her missing the meeting felt like another wrong in a long line of wrongs. But the case I discussed with the leader that day was one of a number that prompted the AGU to form a rapidly moving task force that then rewrote the organization's integrity and ethics policies. They became one of the first scientific societies to recognize sexual harassment as scientific misconduct on the scale of plagiarism or falsification of data, and they now offer free consultation with a legal advisor for people experiencing misconduct such as harassment, bullying, or retaliation. The outcome was not fast enough for the postdoc that year, but it protected her and all of us soon after.

The problems that postdoc had faced made me think with even more urgency about what could be done to protect the careers of graduate students whose advisors had turned against them. Advisors wield tremendous power in the life of an academic. Beyond training them and helping them produce original work worthy of a Ph.D., the advisor is responsible for leading the approval of the student's thesis, and then is expected to write letters of recommendation for jobs even decades later. Often, an advisor leads efforts to give the student awards as their career progresses, or have them considered for editorships or other important community volunteer positions. Within that structure, many kind advisors help build the careers of generations of students. But terrible things happen, too, and when advisors bully or harass their students, those students generally find they have no recourse, and they end up leaving the research world.

At this time I was at Arizona State University as the directorship of the School of Earth and Space Exploration, and as an academic leader, I had both the responsibility and the opportunity to make things a bit better. One small thing I could do was offer to be the primary letter writer for students who had reason to doubt their advisor would write supportive letters. Of course, I could only offer to do this if I knew the student's work, and they were finishing successfully, but this way, I could encourage students to report problems with their advisors and know there was entirely a path forward for them to stay in the field if they wished.

Over the years, you see, I've become radicalized in this area of reporting. Everyone needs to report. As visible as harassers are to their targets, they can be astonishingly invisible to others. If you don't tell your leadership, they may well have no idea, even if that seems impossible to you. Know that even if you report, justice may not be done, or it may not look like justice to you. Report anyway. Know that if you report you may experience repercussions. Report anyway. Without reporting we have no chance for progress, and without reporting, we know that many more people will be harassed by that same person who harassed you. And, harassers often escalate their actions over time. The longer they are allowed to harass, the more people who will suffer.

ONE DAY I RECEIVED AN EMAIL FROM SOME STUDENTS WHO WANTED to meet with me about their instructor. I walked into our conference room the next day for the meeting, and found twelve students waiting for me. More than half the class had come to a meeting to complain, and it was the week before finals: an unprecedented commitment to complaining about the class, in my experience. This instructor was clearly not doing his job. But the next day, when I met with him to discuss the issues and offer help, he denied everything and claimed the complaints were fabricated.

Then, his graduate students came to me, one at a time, and asked for help to find a new advisor; they each felt the training they were receiving was insufficient. I talked with him again about what was happening, and he still refused to take help from me, either in the form of advice and strategizing, or training from internal or external sources. At his contract renewal, which came that same quarter, I voted to retain him because I believed he could turn the corner, but I was overruled by the dean and the provost, who denied him a contract renewal. As it turned out, they made the right decision.

This instructor immediately grieved against me in the faculty senate and sued me in court, claiming I had damaged his career through my racism. Just in case you missed it, I had voted to keep this instructor in his position and support him to turn around his performance and succeed. At our first meeting, the university's general counsel and our vice provost told me their view that this was sexist retribution on the instructor's part. That was when I learned the instructor had not grieved against or sued the dean or the provost, and that I was the only woman in the chain of this process, and I was also the only one who voted in his favor.

In the end, the faculty grievance committee found me blameless for all of the counts he filed against me, and the instructor dropped the lawsuit. But this is the kind of additional stress that women and minorities may experience once they reach leadership positions, because of both implicit and explicit bias.

Meanwhile, a story had broken in the popular press that a senior faculty member in my department had sexually harassed some women at a conference. Then more stories emerged: it was alleged that he'd harassed women at his previous institution. My university immediately opened an investigation, and we began asking publicly at the university for anyone with additional information to please report. And then, suddenly, we had some reports. We'd had none before.

The problem was, the stories in the press were highly inflammatory and we had a group of graduate students who were outraged that they were in the same community he was. I started getting emails from graduate students concerned that they even shared a campus with him. They felt unsafe when they saw him walk by. They did not believe that the university was acting fast enough. They started making demands.

We scheduled meetings for me with all graduate students who wanted to come, so that I could listen and they could all be heard. The first one was coming up in a week. The dean announced that pending the outcome of the investigation, the faculty member accused of harassment was not allowed to come on campus except for brief visits to fetch things from his office, and only if he informed the dean first. I breathed a sigh of relief: surely this would help the graduate students feel safer.

At the time of the meeting I went down to our small lecture hall, and I stood at the front, leaning against a table. I was feeling good, confident and competent. I knew what this was all about, I thought. I'd been there, I'd done this before. Unfortunately, I was wrong. Just one look at the thirty stern and serious faces in front of me told me I was wrong.

"What is the university doing to make sure he is punished and we are safe?" came the first question. I explained, "The university is investigating the claims against him, and they are inviting anyone who has a complaint that hasn't been filed to please file it. We need to know what has happened, and without complaints, we can't know."

"How can we know the university is investigating?!" said a particularly angry woman. "I don't think they are doing anything! And I don't think you are doing anything either!"

"I can tell you from authentic knowledge that the university is investigating. These investigations take a lot of time; they require a lot of interviews with people, maybe some of you. It's a process that has to be thorough because it's the safety and career of young scientists if he is guilty but not stopped, and on the other side, it's his job, or his career, on the line if the accusations we hear in the press are not true," I said, to angry rumblings.

Someone else spoke up, a man whom I had had a good opinion of for his rationality. "Another student reported that you told her that if she couldn't take the environment here she should just leave. Can you explain this?"

"Wow," I said, "that's not something I can imagine myself ever saying. Let me think for a moment about what conversation I've had that could possibly be misinterpreted that way." I was silent for a moment, looking down, wracking my brain for what conversation that might have been. I simultaneously worried that the students might be seeing this as a guilty silence, where I was searching my mind for excuses. And then it suddenly came to me.

"I think I know, yes, I think now I know who this was and what conversation it was. This student was telling me about her unhappiness here, and that she had had happier experiences elsewhere, including her summer internship, which she was continuing as a research project. She was talking about how she might want to spend more time at that other institution while working on the project. I said it was good to have multiple working relationships and institutional connections, and that since things had been rough for her here, working there for a while seemed like a very positive choice. I was not in any way *telling* her to leave. I was supporting her own ideas about her career path and reinforcing that they were reasonable and workable."

The man nodded, sullen-faced, but I do not know if he believed me. What I did know in that moment was that that same deeply unhappy student from the conversation had been creating, whether intentionally or unintentionally, a peer group of disgruntled people and that bitterness and anger was spreading like acid, dissolving the goodwill and sense of community I'd been building up for the couple of years I'd been there.

"But you haven't answered my question—how can we know the university is doing anything?" said the first student. "We don't believe people who are powerful or famous ever actually get investigated. Nothing ever happens. It's always a farce, always the universities protecting themselves."

"Unfortunately," I said, "you're not meant to know what's happening. The process has to be confidential. So what do we do as people who are not in the positions of power doing or organizing the investigation? We wait. And, the university has banned him from campus except for short visits. When he is coming on campus, the dean will let me know, and I will let you know."

"That is not an acceptable answer! I don't feel safe, walking around campus. I never know when I will see him. I don't want to pass him in the hallway! This is not acceptable!" Her voice was rising and her language was strengthening.

"Please, don't shout at me," I said. I was not enjoying this very much. The conversation was beginning to feel personal, and I was beginning to be angry. How could they doubt me, after all the effort I'd put in on making this school safe and productive? And, I thought, wildly, knowing I could not say this out loud, how can you suddenly be afraid of someone in the hallway? He's not going to attack you in the hallway! But then I thought, how do we ever know whether someone is going to attack us? And how can we feel sure, if we've been attacked before? I didn't say anything.

"You can't tell me not to shout at you!" she yelled. "I can tell *you* not to shout at *me* because you are in a position of power above me! But because you have power over me, you *can't tell me not to shout at you!*"

I shook my head, and I looked down at my feet, silent for a moment, trying to keep things from escalating. "I am a human being, just like you. Being in a leadership position does not make me a different kind of person. I have all the same feelings. I work very hard on behalf of this school, and I care about it a lot, and when you shout and accuse me of things, it hurts my feelings, and it offends me." I was laying it out, human and vulnerable, but I didn't yet know how many people were accepting this. Some heads were nodding a little, but most faces were stony.

"Let me tell you," I said, gaining some momentum and keeping the speaking role, "you may never have a stronger advocate for ending harassment and bullying than me. Earlier in my career, I spent significant time trying to get an organization's administration to remove a man who had been accused of being a sexual harasser and a bully. I literally put my career on the line. I've stood up and spoken out when others did not, pressed organizations to change their misconduct rules, protected students with harassing advisors, and worked in groups to stop a man guilty of sexual assault from winning a major award. I really, really care about this stuff. I may be your strongest advocate. And ASU does the best job of addressing these problems of any place I've worked. Nowhere is perfect, but here, we really try to do the right thing."

I paused for breath.

"But how do we know that ASU is actually handling this right? How long do we wait?" asked a new voice.

"That's a great question. Part of our role in the organization is to support the organization to do the right thing. And, if they don't, to stand up and insist. Right now we are in the supporting period. Let's decide what seems reasonable. It's the end of January, and I know the investigation has been going on for at least a few weeks. I'd suggest we wait until the end of February, and if there is no new news, I will ask people up the chain for an update, and I will let you know whatever is appropriate. That might be something like, the investigation is still going on. Let's try supporting what I know to be a good institution trying to do the right thing. Let's support it for now, instead of tearing it down prematurely."

The students seemed placated, though possibly they were just re-signed. And I was left angry and disillusioned. What did they think an attacking mode would get them? When did that posture, going into a meeting, ever bring about a desired result? I wondered if they were that naive, that immature. But dismissing them as childish was too simplistic. And then I remembered a little social media experiment I had run.

In February 2018, I sent out a Twitter poll. I wrote, "Dear Esteemed People in Academia: Have you ever made a formal complaint that someone harassed you? (Know this is not rigorous but truly interested in answers—pls retweet—asking all people of all descriptions.)" Here are the options I offered as responses, and the percentages of selection received from 749 total votes.

- Yes, & result was bad: 12%
- Yes, & result was OK: 7%
- Nope, but could have: 37%
- Nope, never had reason to: 44%

In these results, 56 percent of respondents feel that they have been harassed badly enough that they could have filed, or did file, a com-plaint. And of those, only 12 percent had had an acceptable outcome.

Suddenly, remembering those results, I realized who these gradu-ate student were: They were the *pre-enraged*. They had been harassed, or bullied, or assaulted, and they were filled with the righteous rage of those who had received no reparations. Without that internal drive created by their own past experience they might not have been as sus-picious that the organization was not doing the right thing, and they might not have felt motivated to come to the meeting. And there I was, trying to create that justice, but also with my own history of rage and injustice.

The pre-enraged. This concept focused my thought on the shouting I had heard that day, rage made audible: That rage we carry with us from past wrongs is as real and as hot as any emotion we have. But it is not helpful in creating agreement about equitable paths forward, or correct ethics in leadership, or accountability, or transparency. It's a rare person in leadership who lays it all out transparently in response to the attack of an angry mob. We each need to learn to check that personal rage at the door when we are trying to create institutional change with people who are at least willing to start by listening.

That rage, though, also informs. Remembering the time back in 1995 when I found out accidentally that the man with the same position and arguably less education was making so much more money than I had been, the first thing I do after reaching any leadership position is look at salary equity. I graph the salaries of all employees against some measure of seniority, like the number of years since they received their Ph.D., if they are a faculty member, or against hire year, if they are staff, using different symbols for the men and for the women. In faculty positions, the women are always sedimented along the bottom of the salary range. Then, I'd set about looking at metrics of success in as unbiased a way as possible, and rewarding people who had been under-rewarded.

Similarly, I learned so clearly from the experience with Chris that this is not a fight between men and women, or between men and any other sex or gender identity: This is a fight between people who use power to abuse other people, and the people who get abused. We often just think of women and those with nonbinary gender identities as the recipients of abuse. But men are bullied, and sexually harassed and assaulted, too. As hard as it is for women to report when they have been harassed, I've seen it be as hard or harder for men. They don't think that even the flawed system that exists for women is there for them. One step forward is to talk about this problem. Talk about how men need to report when they are harassed, too. Make sure the paths are known, and announced to be open to all.

We also need to solve the problem of passing harassers from one organization to another. Part of that responsibility lies with the hiring organization. We need to do fierce due diligence. I have started calling references, rather than using letters. On the phone, I can ask, "Given the chance, would you hire this person again?" and they can say no, without putting anything in writing. I've had some memorable phone conversations this way.

One day I was having just such a conversation with a provost at a university near San Francisco. She said to me, "Now when I make the phone calls to previous employers to ask questions like this, I'm always wondering whether I'll find out the person has a history of bullying or harassment. And when other people call me to ask this question, I'll tell them the truth as I know it. Then, I can almost hear the gears going in the other person's head: In five years, if I'm up in front of the Title IV investigative committee and they ask me, Did you have any knowledge of this person's previous history? what would I say? I could only say, yes, I did. And that ends the chain of passing harassers to the next institution, the next group of students."

Chapter 9

CHANGE BEGINS WITH A QUESTION

The room had a long conference table in its center, big enough for the twelve or so students, with more seats along two of the walls, which also carried inviting whiteboards. It was the fall of 2015 at Arizona State University, and I was teaching a one-hour seminar called "Detecting Habitable Planets."

The meeting-room feel of the classroom matched the unusual goals I had for the course: rather than giving lectures to groups of passively listening students, I had in mind a collaborative experience where we decided together where the class would go next. I kept testing my idea against the question, What would this be like in real life? That is, how would we do this if we intrinsically wanted to learn about this topic, as opposed to doing it to earn a grade and some credit?

Some of the undergraduates in the class were looking around a bit nervously. The graduate students—the class was about evenly split between undergraduates and Ph.D. candidates—seemed relaxed, perhaps confident that this would be a seminar-style course. But we were off on a different kind of learning adventure, one that was the result of years of discussion with James and Turner.

That day I gave a twenty-minute lecture on how rocky planets form. Then, I asked for questions. I call these Natural Next Questions—not ones that clarify the content of the lecture, but that instead get us one step closer to the goal of the class.

As expected, the graduate students started off the list. "What did the Earth look like during the Late Heavy Bombardment?" asked one.

The Late Heavy Bombardment was a period about four billion years ago when the Moon and Earth experienced a peak in meteoroid strikes; these impacts created the huge basins on the Moon that are now filled with dark basaltic lava. The students might have been wondering whether the Earth would have appeared habitable then, if viewed from a distance. I'm not sure, because we did not ask questions *about* the questions or critique them in any way; I welcomed and encouraged every one. At that moment everyone in the room needed to feel safe to ask whatever was truly their next question. "Does the creation of a big moon support habitability?" asked another graduate student. And then the first undergraduate: "How did we concentrate water?" The "we" in this sentence refers to the Earth; how did the Earth produce a concentration of water on the surface, so necessary for life? We ended up with nine questions, not quite one from each student, but that did not worry me. The quieter students would speak up eventually. We had enough questions to move forward to our next step.

"Now," I explained, "we will vote. Whichever Natural Next Question wins will be the question we will try to answer for next week's class." We decided that each person got three votes for the questions. We voted, with everyone raising their hands, and we had some agonizing, gratifying moments of "I've already used all my votes! But I want to change one to a vote on this question!" and thus the question to pursue for the next week was selected: Are we an averagely composed system?

"We" in this question refers to our solar system, the combination of our Sun and our planets and small bodies. The phrase "averagely composed" directs us to compare the elemental composition of our solar system to the many other solar systems, systems of planets around stars elsewhere in our Milky Way galaxy, that have been discovered in the past few decades. If the advent of life is dependent upon the composition of the planets, and our system has an average composition, perhaps that means life like ours is more likely to exist elsewhere in our galaxy, as well. Seeing the length of explanation needed for that question, we could immediately suppose that there would be clearer and more complete ways to express it. That knowledge would come with time.

I found a research paper that addressed this question at least in part, and I sent it to the students to read. They had to read it to the best of their ability—we understood and accepted that none of us fully grasp anything we read on the first try—and write a one-page summary of what they could understand from the reading. In the second class, we discussed the paper, and at the end of that class, we made a new list of Natural Next Questions, and voted.

Things were happening in that classroom that I found thrilling. Undergraduates were voicing their opinions even though graduate students were listening. Students were answering each other's questions, building upon one another's ideas. With a little support and prompting from me, the very talkative students were making way for other voices. After a couple of weeks, we decided it was time to sum up all we had learned so far.

I stood at the whiteboard and wrote down "How did rocky planets form?"—the topic of that first mini-lecture I'd delivered—and I drew a box around it. I asked the class, What was the Natural Next Question we'd choose that week? Everyone looked in their notes, and called out the question: Are we an averagely composed system? And I wrote it on the board, and connected it with a line to "How did rocky planets form?" Next, we wrote down the authors and year of the paper we'd read to try to answer that question. Soon, with all the questions and sources on the board in boxes and connected with lines, we had a mind map of the work we'd done so far. I felt pretty pleased! It was already a lot of material.

But then, the most important step came. "Let's talk about how well we've answered these questions. Let's start with that first one, 'Are we an averagely composed system?'" A moment of silence, and then a brave student offered, "It turns out there is high variability in the measured compositions of stars, but at least we are not alone in composition space. There are other stars composed like ours." But then another student said, "It turns out there is a lot of variability between the measurements different labs make. So we don't really know what the 'real' measurement is." The first student jumped back in. "And, it turns out, the composition of the star may not predict the composition of the planets!"

We were able to express, then, as a class, that we didn't think we were alone in composition space (that is, there are other stars out there composed of the same materials our star is), but that the star's composition may have little to do with the composition of the planets, and therefore does not predict whether the planets might be habitable.

"We can't even answer this question!" one person exclaimed. And then, the breakthrough: "The problem we have is that we asked the wrong question. Our question was badly posed. We really wanted to know, Does the composition of a star predict the habitability of its planets? But it turns out we don't know enough to answer that question."

We had a similarly frustrating time with the second and third questions we'd approached, though we did better with some of the later ones. One person suddenly noted that the question in the third week, Are there better kinds of stars for habitability? was just not specific enough, not well-posed enough to answer in a week. "What does 'better' mean?" he asked. There was a moment of stunned silence.

The person who had asked the question in the first place said she'd meant stars that both came from systems with plenty of the elements necessary for life, and also that were not as violent in their radiative outbursts. This led to a back-and-forth over how much we really knew about the elements needed for life, and how we could tell if a given star system had them. One student cited a paper we'd read, and another cited a different paper that had studied a different star system and come to a different conclusion. The data was contradictory.

And then, in a pause, a sophomore said, "How do we *know* when we *know* something?" We all laughed, and then we stopped laughing. We realized that was profoundly difficult to answer. How much information do we need? How do we know if the information is right?

If I could get all of the world to ask themselves that, my work would be done.

From that moment, the students started fine-tuning their questions, and from that moment, I started working on a rubric to score the excellence of questions and guide students to better questions faster. I felt like I was on fire—I was so excited about the idea that we could actually score the productivity of research questions! We were experimenting and making progress together. The class became more interactive and energetic every week. Students told me it was their favorite hour of the week. Some began thinking about unanswered questions that they could actually use as research questions and write papers about.

THIS EXPERIMENTAL CLASS IN 2015 WAS BORN OUT OF AN EPIPHANY we had in a diner, of all places. One day that summer, Turner, James, and I had been eating eggs Benedict at our favorite restaurant, Elmer's

Store, in Ashfield, Massachusetts, and were once again discussing whether we can teach a sense of autonomy to students as they go through what we consider the natural way of learning in the twenty-first century: ask a question, consume some content (internet, anyone?), summarize and critique the information, ask another question. We decided that we could, and we agreed that very morning to start a company, now called Beagle Learning, to build an online platform so we could do this education at scale. Turner stepped in to be the CEO and launch this company and for five years now we've all been working together every week with our cofounder, Carolyn Bickers, attempting to nudge the needle of education toward giving students agency and the confidence and process to solve their own problems.

I've come to think that the traditional ways of teaching science and math are like trying to train dogs by using electric collars. The student is expected to undergo many tests that contain negative corrections (grades and harsh criticisms). I used to think that if I didn't thrive in that environment it meant I was failing, and it meant that I was weaker and less valuable than someone who did. I felt that way at MIT: I was surviving, just surviving, the system of education, being constantly judged and seldom encouraged. I could not imagine a future role for myself in that world, which perhaps explains my years away from academia.

Failing to thrive in that system is not the same as failing. There are some people who thrive and progress in the hierarchical world of corrections and authority. Those of us who do not are not less valuable and we are not weaker. A highly specified, punitive training environment encourages performers who do exactly as they are told, do not stop and think too much, and do not offer new behaviors and ideas. For maximum creativity and breadth of ideas, you've got to get rid of the shock collars and the hair-trigger timing. You have to accept every new, naive, poorly posed question as a valuable contribution, knowing that the student will ask a better one next time. But without the space and encouragement to explore that first question, there will be no next time.

The testing-dominated, hierarchical culture may be a significant component of the lack of diversity in STEM fields: If you don't feel part of the in-group, and therefore have little peer support, then withstanding a reward-and-punishment culture is that much harder. That culture endangers the persistence of women, endangers the persistence of anyone whose primary sense of self is not that of a combatant. Why not go somewhere that feels welcoming and where you have the experience of inclusion and success?

In the United States, learning had come to mean blindly accepting the content delivered in lectures, and returning the same information on multiple-choice and summative tests. I'd bought into this paradigm. In fact, when I was in high school and an undergraduate, I defined myself by it, as many successful students continue to do today. I watched this system winnow people out, either because they really did not learn the material, or because they were just below the curve of the incredibly high standard of an elite university. Or, they simply felt they were failing or they didn't belong, whether or not it was true. I did not question this paradigm; it took moving to ASU and hearing President Michael Crow discuss how excellence can *always* be attained by denying access to 95 percent of people, but that remaining 95 percent make up the majority of the world, and they need access to education and the advantages it confers. He argues for access for everyone ready for college, and he argues for our responsibility for their success.

Questions are a key part of this transformation. Most commonly today, teachers ask questions of the students. A better world, though, is one where students ask questions, where all of us ask questions because we are all students. Sitting in the classroom as an undergraduate, and even as an adult back in graduate school, I really felt that inequity. One day in MIT's class numbered 18.086, Computational Science and Engineering II, Professor Gilbert Strang asked the whole room a question about, I thought, convolutions (the mathematical operation of discovering how one function changes the shape of a second function). I knew the answer, for once, but I had that jolt of fear familiar to many of us when considering answering a question in class. I knew he knew the answer, and that effectively every question he asked the class was a public exam question. Could I get it right, or was I wrong? I raised my hand, and he called on me immediately and I gave my answer, and it was utterly wrong. Apparently he had not said "convolutions" but something else, I don't know what. He paused for a moment in silence, and then called on another student. I felt humiliated and that was the last time I raised my hand in that large class.

Later, in research groups and in seminar-style classes, I experienced the joy of everyone being able to ask questions, and everyone being able to answer if they knew the answer. Having all of us trying to gain as much understanding as possible, all together, felt so much more genuine and also so much more motivating. I began to think about how I could make my classes that way, too. I thought about grades I had received: grades that were lower than I wanted them to be, but felt like some final judgment. I thought about how grades might work instead: Grades could be a conversation, starting with "good job on coming this far, and with all you've achieved. Now here are some next steps that will challenge you and help you progress."

Why do we have to wait until graduate school to learn how to learn? Why was everything presented to undergrads in lectures and meticulously curated textbooks? That's not what learning is in the real world anymore. Information lies around us all in an ocean. We don't need to go, cap in hand, to a university to learn anymore. Instead, we need to be master learners on our own. And we need to do it as undergraduates, rather than only as those few students in graduate school.

My colleague at ASU Evgenya Shkolnik and I started planning a way to create a whole major around the principles of teaching process, and agency, and work-and-life readiness. Evgenya and I had been teaching these inquiry classes together for a few years, and Evgenya was also the principle investigator (or PI, the leader of the mission, the people who sign their name to NASA to make the mission a success or step up to cancel it) of the SPARCS mission, to learn more about planets orbiting M-class stars. What a gift it is to work with another woman leading a space mission, and motivated to try to make a difference in our systems and processes. Her brilliant, rapid brain and ever-ready generous laugh have brought us through many challenging moments in this pathway, but now we have a new undergraduate major, Technological Leadership, as a next step in trying to change the path of education.

Starting with that great class when the sophomore asked, How do we know when we know something? we've been obsessed with questions as the gateways to learning. Questions are more important than answers, in so many ways: answers are almost always incomplete, or come with caveats, or disagreements. To negotiate the ambiguities of answers, we need more questions. Though there may be no such thing as a stupid question, as the old saw goes, there are certainly questions that lead us more directly to our goals and that reveal those ambiguities more clearly.

As we followed the student questions through that semester, and the semesters that followed, we began to think about the characteristics that the best questions held in common. Turner and James and I worked on this, and so did Evgenya, and others who helped us score questions and calibrate rubrics. How do we judge a question to be more productive? What makes a better research question in science, or in politics, or in art?

One evening James and I sat under the mesquite trees in our backyard and scored questions for the very first time. Though he is a mathematician, and doesn't know much of the jargon of planetary science, we each scored a list of questions from my planetary habitability class. We had the hypothesis that expert question-askers like ourselves have some commonsense thoughts on what a great research question is, and that that sense might transcend discipline. We each scored every question for its scale (too small, just right, or too big to answer?) and its articulation (clear, grammatical, specific?). We found, to our joy, that our scores were highly aligned. Of course, there was nothing scientific or rigorous about this experiment, but it gave us both encouragement and ideas for next steps.

As a team, we developed our Question Productivity Index. Scoring relies on the idea that the Natural Next Question you're considering is today's, or this week's, step in pursuit of a bigger goal: a big project goal, or a broad understanding of a topic, for example. A Natural Next Question can also be a research question for creating new knowledge, as in scientific research.

The first part of the three-part Question Productivity Index rubric is Value. How valuable would answering this question be for answering your goal? If answering your Natural Next Question is vital and necessary for answering the big goal in the end, then your Natural Next Question scores five out of five on Value. For lesser degrees of necessity for the goal, the question receives lower scores. In class students have become clearer, using this part of the rubric, over whether they are pursuing their question out of curiosity, or because it really leads to their big goal. Curiosity is great! Following those questions from curiosity can lead us down rabbit holes of specialization, or to exciting personal discoveries that have the joy of a chosen hobby. But it's good to be clear on when you are doing that, and when you are proceeding more efficiently toward your goal.

The second part of the rubric is Scale. Is the Natural Next Question appropriately sized to receive at least a partial answer from reading one article, and to drive a conversation of about an hour? That's what we aim for in class: one step. Many questions can be too big and aspirational; really, they are goal questions in themselves. And others are too small: google them and you have your answer.

The final part of the rubric is Specificity. Would everyone in the group understand this question to mean the same thing? Is the question clear, specific, and actionable? Does it contain no subjective terms or bad grammar? Specificity, it turns out, is key to an answerable question. In that early class when the students realized that to answer the question Are there better kinds of stars for habitability? they had to define "better," a lightbulb went off in my head. And now, we realize that the question is in no way specific enough to be answered clearly. Depending upon the intentions of the asker, questions that would score more highly on Specificity might be, What classes of stars have dynamics and life cycles that would allow nearby water-bearing planets to retain their water for the longest possible time? and What compositions of stars most commonly form planetary systems that include water-bearing rocky planets? And now, week over week, semester over semester, our students get better at asking questions, and they take those skills into their other classes, into the workplace, and into the rest of their lives.

I WAS SITTING ON THE COUCH IN OUR CHILDHOOD HOUSE IN ITHACA and talking with Jim about high school math. I might have been asking him about his advice on what classes to take in my next year in high school; I don't remember the conversation exactly but I remember staring at my math textbook, which was sitting next to a cheap, ornate metal lamp on an old wooden octagonal crate we used to keep our blocks in, and now served as an end table to the couch. I remember the challenge of dusting each of the metal strip scroll decorations on that awful lamp.

Jim said to me, "Well, why don't you just move ahead a year by learning the topic in the summer? You can just teach yourself. Get a book, and work through it, figure it out. It's just high school, after all."

I felt my reality shift, as if I were standing on the top of a cliff and looking over. Wait, just learn it myself, without someone professing it to me and telling me what to do next? Suddenly the world felt open and expansive, and also a little vertiginous. Jim was telling me that high school math was a kind of lowest common denominator, something everyone should be able to learn. Something I should be able to learn, even by myself. He was telling me that my learning was under my own control.

And that is the gift we are trying to give everyone.

As a society we know that education is inadequate; we know that too many students are not learning, that they are not ready for the workforce, that many have never learned to read, or to add. And educational researchers also know how students learn best—in fact, have known for decades through thousands of research studies, though teachers and administrators still do not apply this knowledge throughout the educational system. Students learn best with active, rather than passive, methods. Passive learning is sitting and listening to a lecture. Slightly more active learning would be listening to the lecture and taking notes. One notch higher is discussing the lecture in pairs and then reporting back top conclusions and questions to the class. And the very best kind of learning would be taking the lecture topic and doing your own research to learn the content, and then creating your own lecture. The best way to learn something is to teach it. This is what my brother was offering to me.

As it is, most education is passive, boring, and safe. But not only do we know how to teach better, we also know that the skills of listening passively and regurgitating what we've heard onto an exam are not skills needed for work and life.

In our classes, the students understand from the outset that there are no midterm exams and no final exams, in fact, no exams. So they cannot zone out and snooze and then hit the textbook and pass the course. In fact, there is no textbook.

For grades, if they want to pass, they have to do the weekly work. They get scores for their research summaries, and for their Natural Next Questions, and I also use a "completion score," whereby they get points just for handing in each assignment and equal points for attending each class. That way, I am more freed to give scores based on the rubric for the work, rather than tallying any sympathy points for just having tried.

Then, most important, students get extra credit for improving their scores over the semester. Think about the classic school situation, where a failing grade early in the semester can ruin the grade for the term, and can even discourage a student from continuing to grapple with that topic (think of me giving up biology because of the challenge of organic chemistry). Instead, let's reward them for figuring out their subject over time. After all, as James says, if a chef eventually learns how to make a lemon soufflé, they are now certified in the soufflé. Their previous failed efforts are part of the process and do not count against them. That's how real learning works. Finally, because students are pursuing their own goal questions, they are almost always feeling motivation because of curiosity.

Effectively we are teaching undergraduates the skills of question-asking, research, synthesis, and opinion creation, which are generally only taught to graduate students. But these are skills that every person in our Information Age world needs. Undergraduates can learn these skills. High school students can learn them. And we had better teach these skills, for the future of our species.

At the end of the course, I ask each student to give a four-minute presentation on everything they learned that semester. To fit their summary into four minutes, they need to analyze and synthesize what they have learned into themes; they cannot recite the activities of each week in order. Teaching like this is like x-raying the brains of your students: instead of shoehorning them into an exam structure identical for everyone, you let them show you all they have learned. No two students have the same presentation, no two brains have learned the same thing. Same for the listeners. No two listeners of the same presentation have the same reactions, or remember the same thing as best, or most in need of improvement. Once we get down to this individual level, no two humans think the same way, and communication is necessarily imperfect. At this level, we are not all the same.

We also talk constantly about the learning process. How good was that source of information? How did you find it? How did you think about discovering its conclusions, and recognizing the support for each conclusion? How well did that group discussion work? How could we do it better? Everything is metacognition. If we can analyze why we are doing what we're doing. If we can see what we are missing. If we can imagine how to make a change that improves the structure and process. This is civilization. This is progress.

Chapter 10

ON NOT BEING A HERO

My father had a lot of friends on the faculty at Cornell University, though he himself was not an academic. He was extremely well-read and loved the erudite conversations they would have about literature, or art, or science. One of his good friends in his later years was David, a professor emeritus of geology. My dad wrote me a note one day, in his characteristic all-uppercase block writing, describing a conversation he was enjoying about the nature of fjords, those narrow, mountain-lined ocean inlets characteristic of Scandinavia.

Dad wrote me, "Fjords are so narrow and steep-sided. I wonder if their bottoms, under the water, are flat, or v-shaped? Who better in the world to know but David! I called him right away."

Huh, I thought. Why not me? I was at that time a professor of geology at MIT. Why would it not have crossed my father's mind that his own daughter was a geology expert? He never asked me, and I had no gracious way to enter the conversation, so I just appreciated his sharing and his pleasant conversations, and I was glad he had new exciting friends with whom to share intellectual adventures.

When I was a young girl, my mother used to chide me for having an "earthy" sense of humor; I was supposed to be mannerly. She told me that a girl should never be dominant over a boy. At the same time, she said, she hated being overlooked by her own father in favor of her brothers. My brothers and I, on the other hand, had no approximate equality because I was eight and ten years younger than "the boys," respectively, but we developed a closeness manifested in part by an intensely rapid-fire style of humor. The louder, the more outrageous, the wittier, the faster, the better. One morning when they were both home from college for winter vacation, Jim and I had already "velocitated" to a fever pitch by the time Tom came downstairs and sat at the table for breakfast. He listened to us while gazing with bleary eyes. "Come on, slow down! I just woke up!" he pleaded.

But that level of loud, rapid, pointed humor is not appreciated in many places. Women were not always supposed to be loud and brash, and women were not always supposed to be experts. With my father and the fjords, I realized, he had never seen me as an expert. I began to watch how people who were considered experts spoke, and how they were listened to, and I began to compare those kinds of expert behaviors with those of women who were not already experts. I especially began to analyze how women spoke and were included or excluded in meetings.

I remembered what it had been like out in the field, being flatly contradicted in my observations. Sometimes I was right and sometimes I was wrong, but those moments were always more about dominance than they were about learning. My challengers' thoughts were seldom expressed as questions or counterarguments, but instead as assertions. Of course, it wasn't just me. Senior scientists commonly school younger scientists with flat statements rather than nuanced explanations with support and reasoning, which are usually the required mode of discourse for junior people. I have to resist the temptation to make flat assertions myself, now that I am further in my career. Flat assertions are intellectual laziness, in their avoidance of the hard work of supporting statements, but sometimes we each feel lazy.

I thought about the geophysics group at Brown, where dialogue based on supporting information and questions was more the rule, and everyone learned together. How did they create and maintain that culture?

THE PSYCHE MISSION CAME TO BE BECAUSE OF THE SCIENTIFIC RE-search of my colleagues and me, but to be an effective leader I have needed to learn a lot about the engineering of the spacecraft. I had the good fortune to have worked in those high-pressure labs and learned about building things, and to have earned a forty-hour metal machining certificate, and to have studied during my Ph.D. the properties of materials. That's made it all easier. I ask questions as we go.

Designing and building the spacecraft includes a regular pattern of meetings that start with group meetings about immediate work tasks and ratchet up through reviews of work done by the whole team. Sometimes these reviews are led and judged by a standing review board, hired, curated, and paid by NASA headquarters. Sometimes I serve on panels of reviews that are a step below those NASA-led major-milestone reviews. One week I was serving on one, a review of the Psyche power system, as the scientist, in which my job was to ensure that no engineering decisions were made that would compromise the science of the mission. At the lunch break, a colleague came up to me and said, "What does it feel like to sit all day in a meeting when you have no idea what we are talking about? It must be so boring!" An adrenaline shock ran up my body. I simply replied, "I probably understand a lot more of it than you realize." But his comment—the very offhandedness of it, just making small talk in the lunch line—has haunted me since. His sentiment came from deep-seated bias, bias that made that idea seem natural and normal to the speaker. Would he have assumed a male leader knew nothing about engineering, and made such a remark?

In the end, I realized that all the assertions, and contradictions, and lecturing, and bald statements about another person's ignorance came down to being part of the ego economy. For a millennium, academia has created and curated knowledge using the model of one senior scholar for each discipline, with a pyramid of people and resources under their leadership, a virtual mountain on which they stand, or from which they fall. In this "hero model," research disciplines are siloed off from one another. Each senior professor can choose to protect their hilltop from all comers, and produce new research that adds incremental slivers of real estate around their primary topic. Culture is set in these groups, and so harassment and bullying and all their attendant ills, including a lack of inclusion and diversity, can go unchallenged for decades.

As academics we learn how to deliver ideas in a way that implies we know all the background and are comfortable and confident in our expertise. We speak with periods, not question marks, at the ends of our sentences. Sometimes, we lecture. But this delivery style gets you listened to, and people suspend their doubts and sometimes believe you without further inquiry, if you speak with such authority and confidence.

Thus, the ego economy. Being respected and believed is almost a prerequisite for a successful scientific career. In Germany as early as the sixteenth century professors were expected to earn their success with fame, applause, and charisma; these are not qualities that necessarily align with correctness and expertise. In the worst cases, holding on to that respect and belief, through almost a bully pulpit in your field, can begin to be a preoccupation. Of course, that leads to a hierarchy in how much people are respected and believed, and it leads to a culture where scientific leadership can have their attention focused outward, and begin to groom a fame for themselves that will help to support and advance their scientific careers.

And there you have it: one person's word is law, and another person's word is not heard at all. This is a state antithetical to the best progress of research and though many people and institutions resist it, the structure of scientific research reinforces it.

Years before when I was working at Touche Ross I went with Dan, a partner, to visit a new client, Peter, the founder of a metallurgy company. Peter ushered us into a small conference room with a round table. We all sat down. Peter and Dan faced each other and commenced to have an intense and productive conversation between the two of them. I did not see a moment's opportunity to break into the conversation of my own accord. Though I was the junior person in the room, and a woman among male leaders, I had been brought to this meeting for a reason: I know something about metallurgy. I did not get an opportunity to speak, that afternoon.

Somewhere along the way I began to notice when a group did work the way it should: Every person with something to contribute had their moment, and everyone else listened. Every person was valued for their own expertise. It was so relaxing, when it happened. I saw the same thing happen in some early meetings on the Psyche mission. Everyone who had something pertinent to say had their moment. Soon we had our unofficial team motto: The best news is bad news brought early.

SINCE COLLEGE I HAVE BEEN PERIODICALLY ACTIVE IN QUAKER MEETings. Turner had a lovely baby welcoming ceremony—similar in some ways to a baptism—in the Quaker meeting in Annapolis, Maryland. The ceremony was held in a little apple orchard and meadow where, when there was enough money, the meeting later built its own meetinghouse. That day was filled with the clean smell of sweet spring grass. The Quakers believe that each of us has our own relationship with the divine, and that no one is needed to stand between or interpret. Rather than ministers, these Quaker meetings have clerks, to keep administrative order, but not to dictate to you your own spiritual experience.

Later, James and I were married in the Quaker meeting in Acton, Massachusetts. For some months ahead of time we met, apart and together, with a clearness committee. The committee's task was not to dictate whether or not we could be married, or to school us in doctrine, but to help us determine whether we were entirely clear in our thinking and our decision to marry. These aspects of practice appealed to my unwillingness to be given orders, and calmed my fears of having unacceptably mismatched beliefs with the Quakers.

The Quakers have made efforts for centuries to create a more equitable world. Quakers were the first to invent price tags, since it seemed inequitable that different people would pay different amounts for the same goods. In a Quaker business meeting the practice is that no one speaks twice until everyone has spoken once. Why not apply the same concept to my own teams as well as in the classroom? I thought.

I began to build a set of concepts to guide team building:

- Leaders, speak often about the culture you want.
- There are reasons why any given person might not feel able to speak up; invite everyone to speak, and clear your mind of preconceptions over whether that person has value to add. Listen to every voice.
- Believe everyone comes with best intentions.
- Learn to be truly happy for others' successes.
- Don't create boundaries—keep everyone on one team. Boundaries around sub-teams make people act like inhabitants of small villages: Each competes against the other, and in fact, views them as "other," not as a teammate. Subdisciplinary boundaries within academic units are particularly damaging.

- Bring problems and challenges to the whole team and invite people to solve them with you. You need not solve problems in isolation.
- Treat everything as an experiment—failures are okay. The object is not to outlaw mistakes. That should never happen and it's counterproductive, as it makes people nervous. As NASA Associate Administrator Thomas Zurbuchen says, "The goal is to catch all the mistakes. We'll never as humans be able to stop making mistakes, but with a team, we can catch them."

I had begun to form my own ideas about how teams should run, but in academia, I was still surrounded by the hero model. The hero is the single loudest voice and the arbiter and doer of all things. The hero feels unattainable to others, and the idea of a hero is a fallacy. No single person can alone build human knowledge anymore. We need the breadth of ideas that comes from a diversity of voices, and we need to move faster than is allowed by most people's single next idea. Making real change or real progress is difficult with a team that thinks each person's reputation is the most important thing. I wanted to bring people together not around an ego, but around a question.

And then, in aerospace, I saw another form of hero: the aerospace cavalry. When a flight project is late and errors have not been corrected and the launch date is approaching and everything is on the line, these organizations can mobilize an engineering attack team that comes rolling over the hill like the galloping cavalry. They move in and seize the project from the existing engineers and work heroic hours and pull heroic wins out from seemingly certain losses. One example of the aerospace cavalry occurred with a recent Mars mission. The team building one of the instruments was having trouble successfully building the hardware, and as the test failures and parts breakages mounted up, the schedule slipped further and further behind. The costs of the extra work and extra hardware and the strain on the rest of the mission made a sound like coins clattering down the drain. In came the cavalry! One of the NASA centers assembled, at significant additional cost, a team of the world's top experts on this kind of instrument. This cavalry moved to the site of the instrument build and, effectively shouldering aside the people who were failing at the build, just made it happen themselves.

Every one of the aerospace cavalry is a hero: it's hero culture times one hundred. The heroes save the day! But it takes a huge toll, in exhaustion, in money, and in a sense of failure of other team members. So I decided the goal was never to be the hero and never to have to call in the heroes. The goal is to do it better from the very beginning.

THERE WERE FIFTY PEOPLE IN THE LECTURE HALL AT ASU THAT DAY in late January 2017 as I prepared to test-drive this new leadership technique. I was standing at the front, looking across the auditorium seating, so relieved that people had come to this inaugural meeting for the Interplanetary Initiative that Michael Crow and I were cochairing. The curving ranks of auditorium chairs were in front of me as I stood under the somewhat harsh fluorescent lights with a whole wall of empty blank whiteboards behind me. I'd invited people from across the ASU campus and our broader community. As I'd watched the coffee and pastries being set up by catering, and checked on the whiteboard markers and poster paper, I'd had that anxious pre-party feeling: Will anyone show up? Now that they had, I turned to my next concern: How would we fill this veritable airport runway of whiteboards with ideas?

Once people settled, we went around the room doing quick introductions: The dean of the College of Integrative Sciences and Arts. A graduate student from electrical engineering. A professor of psychology. An undergraduate studying Earth science. A researcher in public health. A local philanthropist.

"We spend too much time," I said after everyone had a chance to say who they were, "as academic researchers, answering the most obvious next question, the one that extends what we've already done one small step, or applies our idea to a new context. We spend too much time being incremental. We don't have the time or the money to be incremental. Today we will practice transformation instead of incrementalism."

What, I then asked everyone, are the biggest goal questions we need to answer to create a positive human space future? For this was the central motivation of the Interplanetary Initiative: to bring together all disciplines—not just engineering, science, and visionary sci-fi writers—to create our space future. We wanted the sociologists and psychologists and business leaders and policy experts, so that we could make faster and more targeted progress than university initiatives generally do. And we wanted our projects to connect with society, to invite all of us to look up from our dusty feet, and see our places in the universe. Heady stuff.

I wasn't after the questions any one of us might answer on our own, or with our lab, or in the next year. We needed to find the questions that were truly the keys to the positive future. "Imagine," I told the group, "that the questions are javelins you've thrown out so far they land over the next hill. Our job with this initiative is to set up research teams to run after them, to see how close to answering them we can get."

First the questions came from a few confident faculty members. Then the cadence sped up—more and more people had ideas they wanted on those boards. Students spoke up. People from the community spoke up. In proper brainstorming fashion, I wrote down without criticism or editing every question that people called out. Our list grew:

- How do we make the space experience and data available to the public?
- What near-Earth satellites and missions are needed to improve life on Earth?
- Can space be a unifying venture rather than a proxy for war?
- How can we better connect robotic and human exploration?
- When there is a colony on Mars, will it unite or fragment Earth society?
- What does past exploration teach us about future exploration? How do we fix our past mistakes?
- How will Earth react to the first baby born in space?
- What legal, political, and social norms will govern human exploration?
- How will humankind react to the discovery of life off Earth?
- How do we prepare for psychological safety in space?
- Will we all be humans in the future?
- What kinds of social structures will be required for survival in space?
- Can you have democracy in space?
- How do we make the space narrative harmonious with sustainability on Earth?
- How do we "breed," educate, and prepare a new generation of explorers?
- What is the business model for interplanetary exploration?
- How do we stay connected across the distance?
- How do we be inclusive while exploring space, and prevent widening gaps between nations?
- How will multination, multiworld human society be governed?
- How do we galvanize public/private support to make this happen?
- How do we open our technological process to greater leaps and disruptions?

- What steps in the development of space hardware need modern user experience design?
- Why are we doing this?

Across the entire front of the lecture hall, the boards were filled with questions. People sat back, and there was a moment of quiet.

"Now," I said, "we vote."

I was pushing this group, mostly traditional academics, well out of their comfort zones. I kept making eye contact with everyone I could around the room. I wanted that real human connection to keep them in the game. Many people met my eye and smiled, but some sat back in their seats and I worried they were formulating an internal critique. I was on a tightrope; I feared that if a few people started making disparaging remarks, or walked out, my social capital would be spent and I'd have a hard time starting up our research program. These brainstorming processes are common in start-ups, common in business, but rare in academia. In academia people often keep their ideas and questions to themselves, and voting on a research direction is pretty much unheard-of.

We used the same three-vote method I had used with my class. Eight questions rose to the top:

- How do we make the space experience and data available to the public?
- How can we better connect robotic and human exploration?
- What steps in the development of space hardware need modern user experience design?
- What does past exploration teach us about future exploration? How do we fix our past mistakes?
- What legal, political, and social norms will govern human exploration?

- How do we make the space narrative harmonious with sustainability on Earth?
- How do we "breed," educate, and prepare a new generation of explorers?
- What is the business model for interplanetary exploration?

By now, we'd been at the work for almost two hours, but there were two more hours to go. People were expectant, awaiting the next step. Very few had left the lecture hall.

"Now," I told the remaining group, "we make our teams. We're all going to go into the next room, where there are tables, each with one of our big questions written on poster paper on its center. Each person should sit at a table with the big question they are most interested in. Each of these volunteer teams will have one hour to decide the following: What milestones can we reach in one year, in pursuit of this big question? What specialties or disciplines do we need in team members that we do not now have? And finally, how will you incorporate this knowledge discovery into your classrooms?"

People stood, chatting in groups, and began to make for the door. I stood back, holding a sense of growing panic in my belly—would they all just slip down the stairs and out of the building? How many people would continue this bizarre experiment in the next room? Just me and the crickets? But my moment of panic passed as I saw people move into the next room and settle at tables. We'd passed a hurdle. Now, would these new groups be able to create?

Each of the tables was filled. Conversation was loud. I circulated, listening for a while at each table. Group 6 was discussing the question, How do we prepare the next generation for space? They were talking about the importance of risk-taking, adaptability, and tolerance for ambiguity, none of which are taught in the standard American curriculum. Group 1 was discussing, Can we develop a mission concept combining human/robot integration? The students, staff, and faculty at that table were talking about bringing designers—architects, artists, psychologists—together to identify core needs without the mental constraints of what exists today. And Group 7 was excitedly at work on what social structures and practices were necessary for sustaining a social unit for an indefinite time in space. They were talking about developing a set of values that would govern how people will interact in space. They started discussing creating a game to see how people would act, and how their actions could be influenced.

That day Michael Crow came to see how the teams fared. Each stood to present their proposal, how they would involve students, and what other team members they needed. For some teams, faculty stood and presented. For others, staff. And one team was ably represented by a freshman. That night, I slept the sleep of the validated.

In the following weeks, a few of these teams didn't quite make it to the planning stage where they could receive seed funding. Others forged ahead, and we gave several of them seed funding by May. I thought that now the semester was over, we wouldn't hear from them until the fall. But I was wrong: to our intense excitement, we found these teams to be singularly motivated and interested in their bigger, interdisciplinary projects, and by the following September, we had our first peer-reviewed paper published. That paper addressed the question: How will humans react when life off this planet is discovered? A psychologist and his team analyzed the public responses to press releases through the years that announced the discovery of extraterrestrial life. (The responses, that is, in the interval before each press release's "discovery" was retracted as new data invalidated it. As of this writing, we have not discovered life off the Earth.) As long as that life was not big, smart, or equipped with fangs, the press releases indicated, we were pretty happy, interested, and positively inclined about this complete paradigm shift in human existence.

Group 7, which had hatched the idea for a game investigating how humans will interact off Earth, gained momentum rapidly. Their leader, Lance Gharavi, assembled a team of professional game designers, artists, and social scientists, and they invented a game called Port of Mars. Settlements on Mars need to act together to survive. In his Twitter post about their first game test, he announced the success of the test, though the outcome for the putative settlers was grim: "Everyone died." In the intervening years they have published peer-reviewed papers on the data they've gathered with the game, and launched an annual tournament called Mars Madness, and provided a 200 percent return on our seed funding with external grants. In fact, in the first four years, the seed-funded pilots have created a return on our seed investment of 800 percent.

One of the teams set off as a partnership between ASU and a small external company, but the ASU lead faded out of participation and we were left with an external lead only. Financial management became difficult, as did hiring and purchasing for the project. With the challenges this produced, we had one of our early lessons: We always need a strong ASU lead, no matter how good our external partners are. Now, we pick an ASU leader at the end of the Big Questions brainstorming session and we give the teams two or three weeks to come back with milestones, outcomes, and a budget for one year. This, it turns out, is an adequate test of leadership, and it's helping us launch stronger teams.

From the beginning, we've also stretched the usual comfort zone of academics by immediately placing each team under project management. Our project manager, Abigail Weibel, has embraced the experimental nature of it all, and we constantly reassess what is working and how to improve processes, year after year, so that more teams are successful.

Of the twenty-six pilot projects we launched in our first four years, by the end of the fourth year, ten had launched successfully to completely independent funding. Another eleven were ongoing, continuing to make progress. Only five had had their team dissolve. We'd developed a way to drive a project from a Big Question, and not from the hero model; we'd created a transparent way to assemble a functioning interdisciplinary team; we'd made a model that successfully moves from seed funding to external funding.

Some of these ideas about letting go of the hero model for academia came out in an essay I wrote for *Issues in Science and Technology*. A colleague messaged me to say that he'd been indoctrinated into the hero model in graduate school, and embraced it until he became chair of his department. At that moment, he saw it was all wrong. And I'd had a similar growth. Those hero-model pyramids of teams and resources can be opaque to colleagues, but as the chair, you see more of what is really happening. The chair can see—I saw—whose students were suffering, whose work was adding substantially to human knowledge, who was supporting junior colleagues, and who was not. He wrote that describing academia this way made so much clear that it felt like a dam bursting.

Chapter 11

EVERY DAY, A BRICK

I can't know how I would react if Psyche were canceled now, as I write before launch. I can't know how I would react if it failed on the way to the asteroid. What would picking up and carrying on feel like? Perhaps I would feel a little bit as I did in September 2014, when we were writing our Step 1 proposal, competing with twenty-eight other teams, and when after a routine surgery my doctor called to tell me I had ovarian cancer.

Six weeks earlier James and I had moved to Arizona and I had started my new job as the director of the School of Earth and Space Exploration at Arizona State University. I was diving into this big complex beast of a job, with 60 faculty to support, almost 500 students, 350 people on payroll, and operations spread across 7 buildings. Every day was a blur of meetings, listening to faculty, learning the business processes, getting to know the staff who kept it all running. And I had a little lump in my belly that worried me a bit, when I had a moment to think about it, when I was lying on my back in bed and could feel it under my hand.

I had made a doctor's appointment as soon as I could after arriving in Arizona, setting up our house, moving into my new office, and bursting out of the starting gate with the new job. I wanted the doctor to help me with some terrible sciatica that was hindering my control of my left leg, making walking difficult, and also producing some shocking pain. The doctor said, "Yes, we can work on the sciatica, but right now, I'm much more concerned about that lump in your belly. Let's get an ultrasound." The sciatica would have to wait its turn.

After the ultrasound, the doctor said I had an ovarian cyst, which is common. They often resolve without intervention, but this one was a bit large and he thought it should come out. He pointed at the ultrasound. "Only one or two percent of large ovarian cysts are cancerous," he said. "Sometimes when there are interference patterns in the fluid inside the cyst, as you see here," he said, pointing at a checkerboard of faint ripples, "there is a correlation with cancer." I heard him, but without any reaction of my heart or blood pressure, without any dropping of my stomach, because having cancer seemed out of the question, wildly improbable. I went home feeling calm and resigned to having the cyst out. The thing I dreaded most was getting the IV in my hand.

ON THE DAY OF THE SURGERY JAMES AND I WAITED IN A PRIVATE room for the surgical team. I was all changed and propped up in a hospital bed with warmed blankets tucked around me. Oh, the luxury of warmed blankets! The nurses and anesthesiologist listened to my story of exceptional nausea and vomiting after previous surgeries, and they suggested drug adjustments that would help. How privileged I am, I thought, to have this level of care.

Having the IV put in was a predictable horror: The experienced nurse wanted to have the trainee practice on me, and she narrated every step, discussing valves in the veins, and the angle, and the puncture, and I thought I was going to faint, and I thought I was going to throw up. Eventually it got done and I drifted off. Soon enough I was waking up in recovery, not even too nauseated. The surgeon said the cyst came out cleanly and they saw no evidence of cancer. He also said I'd had a tremendous case of endometriosis, and that he'd had to work an unexpected several hours longer than planned in the surgery, removing all the uterine tissue that was growing throughout my abdomen outside the uterus (James had been alarmed, sitting in the waiting room, as the surgery stretched from the scheduled hour or so to four or five). The surgeon commented that I must have had significant pain and problem periods, and that this disease must have been my greatest issue. But in truth, I had not had unusual problems. I did wonder later if endometriosis had made it harder for me to get pregnant.

I went home to recover. But three days later, the surgeon called me. They had found cancer, he said, in the cyst in the postoperative examination of the samples. I remember holding the phone to my ear. I remember the surgeon telling me I had to come back immediately for another surgery, this one including a hysterectomy, and removal of both my ovaries, and biopsies of lymph nodes to discover if the cancer had already spread. I told James the news and we stood looking at each other, steeling ourselves for this next part of the journey.

I called Turner. James was listening as I gave Turner the news on speakerphone, so we felt we were together. Turner's voice was calm. I knew I was experiencing something life-changing, and I kept checking myself to see if I was panicking, or overreacting, or acting. No, I was not. Was I in a kind of frozen suspended animation of emotions? I didn't think so. I felt hyperaware of everything happening and everything being said, but I also felt matter-of-fact and ready for the challenges ahead. My ears were ringing and all the colors in my range of vision looked too bright. But I was still there, fully engaged.

I knew that the five-year survival rate for ovarian cancers broadly was only about 40 percent. The cancer in my cyst was early stage and not as aggressive as it might have been, and so I knew my odds were much better than the average. I focused on the idea that I was going to be fine. Being fine felt like more than an idea: it was clearly the future. I was going to be fine. Turner was calm, too. I could not hear in his voice whether he was covering up any alarm, whether he was just accepting of what would come, or whether he didn't feel connected enough to really react. I told Turner and James, and I felt strongly, that if I had to fight cancer, we would give it the best, most optimistic, most determined fight ever. Turner laughed. We all agreed.

Crisis is a state I understand. Having seen myself react over and over through the years to big events with calmness and renewed focus, I believe this comes from my earliest abuse: crisis was a way of life in those young years, and I was determined to survive it. Crisis now has a familiarity. Its familiarity is almost comforting. No more dreading the worst thing: the worst thing was here, and now we could contend with it out in the open. The Greek Stoic philosopher Epictetus wrote, "To the rational being only the irrational is unendurable, but the rational is endurable." For me, then, a known crisis with specific steps that need to be taken is rational, and endurable.

Knowing that all would probably be fine, but also knowing that I could have a significant diminishment of life quality, and also that I could die, I examined myself for regrets and panics about my life to date. I found in that moment the conviction that the vision and meaning we put into our lives mattered. Every unit of work, every day of effort, was a real piece of progress and of value in itself, and I did not feel I had wasted any time. So I went on trying not to waste any time.

After the second surgery I set to healing again, and got back to work at ASU. I also got to know my fantastic oncologist. He gave me the only two published medical journal studies on the exact kind and stage of cancer I had. At a time when I was still exhausted from the two abdominal surgeries and my attempts to get work done during long days, Sara Collina read the new suite of research papers and summarized the statistics and findings for me, surely one of the kindest and most honest acts of friendship. These couple of studies showed that my chances were probably better if I had chemotherapy. And they showed that if the cancer recurred, I had very little chance of survival.

Chemotherapy affects each person differently. The one I had can have few side effects, or it can begin dissolving the myelin sheath that guards all the nerves in your body, and it can give you foggy, confused thought, and exhaustion. That's what happened to me. Every week that I went to the Mayo Clinic and had those chemicals dripped into the port they had installed in my arm, the hours that I could think clearly and had the energy to work grew shorter and shorter. But in those hours, I was determined. I would get my work done.

James supported me entirely, with all he had. He shopped and cooked. I went to work every day. For a few days I drove myself, but then one day I had a reaction to prochlorperazine, the medication I was taking to combat chemotherapy-induced nausea. Sitting in my office, I began to feel distanced from the objects around me, an almost psychedelic effect of being cut off from the real world and backed away from it. My arms felt suddenly weak. I could barely raise them up to my keyboard. My jaw was tight. I tried speaking a few words and they came out slurred.

I called the nurse at Mayo. She actually berated me for attempting to work while being treated for cancer. She said I should stay home and rest. That left me not only alarmed for my condition, as I had already been, but also furious. The next meeting in my office was with Scott Parazynski, an astronaut who is also a medical doctor. That excellent man asked me how I was feeling, as he came in to chat about some work matters. I told him, and he said, "Oh, that reaction is called dystonia. It happens in a few percent of people who take that drug; it happened to my daughter one time. Dystonia can manifest as you had it, as a partial paralysis of muscles. It'll pass soon. You might think about taking it easy for the rest of the day." Much relieved, I thanked him and we went on to talk about the business at hand. The side effects passed harmlessly after a couple of hours, but for many months after that day, James drove me to work.

My brother Jim and sister-in-law, Margaret, stayed with me for a week, and then Sara, and finally my friend Andie and her husband, Tim, and Turner, allowing James some rest and the chance to do some travel for his work. Coming to care for someone for a week is a huge kindness. I was so fortunate to have so many friends and family willing to do it for me. At work, my assistant, Cami Skiba, and the school director of research and operations, Teresa Robinette, supported me and believed in me even though we'd only known each other a few months. The kindness of my friends and family gave me the ability to keep these two critical projects going: my new directorship at ASU, and the Psyche mission.

In those few hours per day of clarity, I led the writing of the Psyche mission science section and worked with the whole team on our 218-page Psyche Step 1 proposal to NASA. The professional artwork on the cover showed the moment of impact between two planetesimals, with brilliant light exploding outward as the heat of the molten interiors was revealed to space. After a page of stern warnings about export control on the technological information, and the need to limit distribution, two pages of graphics told the story of the mission, from science to spacecraft to the science team. The proposal text started with the science investigations we proposed, and then supported their feasibility with descriptions of the instruments we'd fly to make the key measurements, how the spacecraft would be operated, how much data we could send back, what orbits we'd assume around the asteroid, the systems engineering approach to designing and building the spacecraft, the new technologies we'd need, the management approach for the team, and then a long section on what it would cost and how we knew. I still get a strong emotional response in my chest when I open this document.

We handed the proposal to NASA in January 2015, when I was at my peak of sickened exhaustion, with numbed hands and feet, sciatica reignited by the chemotherapy that further stripped the damaged nerves. The drive to work on the mission and make the new job a success, to support the people in the school at ASU to do their own best work, kept me moving forward through that ridiculous level of physical discomfort. If I needed to roll over at night, I had to wake up and roll myself over using my hands; getting in or out of bed was a torment of pain in my back and down my left leg.

I aged seemingly overnight. Some days I felt I was at my limit just getting dressed in the morning; tears of exhaustion would leak down the makeup I'd just applied over my pallor, over my cheeks swollen by steroids. I held on to a conviction that none of this was forever. I had to believe better days were ahead; that was the only possible reason for persisting through this tangle of daily impossibilities. My hair was gone. Seeing myself puffy and hairless in the mirror was surprisingly upsetting. The idea of losing my hair temporarily had seemed such a trivial problem. The shock, however, was failing to recognize my own face.

IN SEPTEMBER 2015, AS MY STRENGTH WAS COMING BACK, THE CALL came: we were one of the five teams selected to go on and compete through Step 2. We were ecstatic, the kind of creamy, unalloyed joy that only comes with surprise.

And this wasn't that fleeting, equivocal happiness that comes with winning a prize, when the next day you're the same person, no new goal, and no real change in your soul. We experienced a lasting joy because this win was not an empty ending, but a thrilling beginning, like being told you were so good you were being specially selected to run an exclusive marathon. Now the starter gun bangs, and we're off. We had a new, giant, lengthy challenge to pit our courageous selves against. This was a kind of winning worth having.

Several of us were new to the NASA process. At the kickoff meeting in November, comfortable in our naivete, we asked, "What happens in Step 2?"

Brent Sherwood, program manager for solar system science mission formulation at the Jet Propulsion Laboratory, explained, "In the next nine months, up through August, we'll respond to all the concerns that reviewers give us in response to our Step 1 proposal, and we'll write the Concept Study Report." Brent, perhaps the most focused and precise strategic thinker I know, and serious and tough as a bulldog, had brought our mission concept along from the earliest ideas in 2011.

What is the Concept Study Report? we asked.

Gentry Lee, JPL's chief engineer for the Planetary Flight Systems Directorate, snorted. Gentry knows all the details of every planetary mission JPL has ever flown, it seems. He had written books with Arthur C. Clarke and collaborated with Carl Sagan. He comes with a baseball cap and a beaming open smile or a foreboding frown, never with neutrality. When he speaks, it's with drama and authority, and he'd become a special friend and supporter of the Psyche mission. He answered, "It's the proposal you've written already, but with every single detail worked out and supported. It's the Step 1 proposal, only an order of magnitude bigger."

And then we'd have the site visit. The serious faces around the table were conveying something I didn't understand. *Site visit?*

"After we hand in our Concept Study Report," Brent explained, "we'll present to NASA's professional review committee. We'll have to start rehearsing early."

Gentry agreed. Since the visit would be in November of the following year, he suggested we start planning in just a few weeks. David Oh, our lead systems engineer and also capture lead (the person on the team in charge of making sure we were as successful in competition as possible), and I looked at each other. Neither of us had been through this process before. Why would we begin rehearsing in the winter for a presentation the next fall? Our presentations about the mission were already polished.

The first thing that happens, Gentry told us, was that we would receive an encrypted email from NASA on the morning of Day 1. This email would contain a list of questions we would need to answer for the review panel, along with instructions on whether the answers should be submitted in written form, presented in person at the end of the week, or both. Then we'd have five days to answer the questions. David and I looked at each other again, smiling. Five days! We could answer anything about the mission in five days.

Then we realized these would be questions that had not been answered in our 218-page Step 1 proposal, nor in the Concept Study Report, which would eventually run 1,053 pages. These were the hard questions.

We had received $3 million for our Step 2 process. Brent looked at me, again unsmiling, and said, "That is not nearly enough. We'll need at least five million to do what we need to do."

David and I had received our first taste of what this year would be.

JPL GAVE US A "WAR ROOM" WHERE THE TEAM COULD GATHER TO work. Down a hallway, through a door, up some stairs, end of the next hall, and there was the door to our windowless war room. A combination lock allowed access only to members of the team, since everything we put up all over the walls and wrote on the whiteboard was "competition sensitive." We had a big conference table that always had people around it working on laptops, and we had a printer from which pages of the latest draft were always issuing, and we had snacks, including a dangerous amount of candy.

Every day, all day, for the whole year, there were people working together in that room. We tacked up all the draft pages of the Concept Study Report on those bulletin board walls, starting at the top left of one wall, and then placing page after page as if we were tiling the room with the proposal. New drafts were tacked on top of old ones; the progress was recorded that way, and changes could be interrogated by multiple people in real time.

The JPL program office pored through NASA's proposal requirements and handed us a detailed table of contents that covered many pages. Different sub-teams started work on different parts of the proposal. David set up the engineering teams to run some models and begin writing the sections on building the spacecraft. Then, he and Brent and I started "storyboarding" the science section. We told and retold each other the science story of the mission until we had some idea of which images and figures were needed to explain the whole arc. We printed out drafts and tacked them on the wall, and reorganized them until they told the story right. I began writing drafts.

We planned the proposal sections in multiple whiteboarding sessions, writing down our primary claims for each section and what supported our claims. For the science instruments, for example, we claimed that three instruments were the correct choices for answering our science questions, and we backed that up with an analysis of what measurements were required to answer each science objective, and what instrument was needed for each measurement, and how the specific chosen instruments would be able to make each measurement, and then, how much data they would create, and how long it would take to download that data to Earth, and how much redundancy in measurement was needed, and how much margin. And then, how many people on the team would be building each instrument, and what their experience was, and how the schedule came together for each one. And so on for a couple of dozen topics of equal complexity.

I worked remotely from Phoenix every day, and then I flew in every week or every other week to join the team in person. And we began to plan for the site visit. I hired a professional speech coach to come work with the whole team. We watched carefully while people practiced speaking, and we used their performance to decide who would speak on site-visit day and who perhaps should not. Speakers learned to make eye contact with the audience, and to use open body gestures, and to organize our ideas into three main points that we reiterated. We learned to "hit our mark," to walk to a mark on the floor where the lighting and viewing was best, and give our talk from there, and we all bought better clothes. We strategized over what parts of the proposal were likely to garner the most concern, and we focused on the people who were experts in those areas. And we scheduled a mock site visit.

Six weeks before the real site visit, we all moved into hotels in Palo Alto, California, where the actual site visit would be, at the headquarters of our partner, Space Systems/Loral (now Maxar Technologies), who would build the spacecraft chassis and power system, should we win. We wanted our reviewers to see how impressive Loral was, with its giant high bay filled with satellites being built, because Loral focused on Earth orbiters and had never had a deep space mission before, and thus was a new potential NASA partner. The management at Jet Propulsion Lab hired a review panel. On Day 1 of our week, we received an email with questions. We set up a spreadsheet, and triaged the questions, and assigned leads and seconds for each question, and made a team schedule for check-ins, and the forty of us prepared our answers, just as we would for the real site visit.

On Day 7 we presented to the mock review panel. The panel was confrontational and baited us, and we did our best to be clear and calm. One panel member challenged me, "What kind of leadership experience do you have that prepares you for this job?" and I responded confidently with my leadership at Carnegie and at ASU and my business experience, but the panelist was not satisfied, and shot back, "None of that is relevant to what you'd do here. Here there is more money, more pressure, and you've never delivered spaceflight hardware. How can you make up for those lacks?" As much as I compared budget numbers from my departments to NASA Discovery mission numbers, and touted the complementary skills of the team, the panel continued its attack until the questions felt entirely personal—we'd moved to the *ad hominem* from the objective. I began to realize that their goal was to beat us up, not support us to practice presenting. The feel in the room was negative. The comments were negative. We ended the practice feeling awful. In six weeks, we had to do it again, this time for real.

THE WEEK BEFORE THE SITE VISIT, MAXAR REPAINTED ITS EXECUtive lunchroom, and hung new Psyche art on its walls. Then forty of us moved up to Palo Alto for the week of the actual event. We had a giant banner of the cover art of our Concept Study Report hung down the whole front of the building, making quite the impressive spectacle from the street. We agonized over the seating chart, and we adjusted the window blinds so no stray light would hit anyone's face. And then, on the morning of Day 1, Monday, November 1, 2016, we got the encrypted email filled with questions.

The questions were unexpected. We had thought that the instrument measuring Psyche's composition, the Gamma Ray and Neutron Spectrometer, would get all the science questions. David Lawrence, its lead, was there and ready to go. But instead all the questions came in for the imagers (the telescopic cameras) and the magnetometer (which would measure the existence of a magnetic field). We told Ben

Weiss, lead of the magnetometer investigation, that he was going to present. Surprise! We called Jim Bell, the imagers lead, who changed his plans for the week and hopped on a hastily booked flight to arrive early the next day.

The forty of us sat each morning to see how progress for each answer was coming. We used the processes and organizational tools that we'd developed in our mock site visit, and we tried to leave behind the terrible sense of failure that that event had left us with. We made models, did research, and passed our drafts to our first and second readers. In the evenings, we went back to our hotels, sometimes eating dinner together, other times working, and sometimes going for a walk to reduce the stress. Each day, more of the questions were answered, and then passed to the documentarians, who copyedited and formatted the material so it was perfect.

We knew that our chances of winning lay largely upon our performance at the review on the final day. Millions of dollars rode upon that one presentation. Not just the millions it took for the year of Step 2, but the hundreds of millions ahead if we won, the jobs for hundreds of people for years. We felt it, every penny.

On Day 6, the day before the panel arrived, Maxar oiled every single chair in the room so that none would creak. They tested every power outlet. We had tape on the floor so we'd hit our marks. We reviewed the advice of our professional speaking coach. The lighting was tested. The videos worked. The PowerPoint presentations were all ready to go. And now the on-site team numbered 140, because we needed a "Red Team," with duplicate expertise to the leadership team, to work on answering new questions that came up during the presentations. We even had runners to go between the Red Team room and the presentation room. Everyone knew their place and their job.

Henry Stone, the Psyche JPL project manager, and David Oh each spoke to the team and expressed their gratitude and their wonder at where we had come. I listened to them and thought how fortunate I was to work with them. Then I stood where I would stand the next day in front of the panel, and I looked out at the 140 people gathered, some sitting, some standing along the windows and in the back.

"Very few people get to experience what we are experiencing," I said. "Very few teams come together around concepts like the Psyche mission, and very few teams get to the advanced decision point we are in. So few people get to go through a challenge and a trial like a major NASA site visit, a life-changing experience, an experience that makes us stronger and better and drives us to understand more deeply what quality is. We are sharing a moment of immense privilege and immense pressure: We have worked together to try to make a perfect thing, a plan to explore a kind of body that humans have never seen, using technology that is the culmination of thousands of years of human effort. We have made the decisions that we knew were right, even if they were not always the popular or expected decisions.

"We have together created a story and a plan that is the best it can be. We need to feel this, to live in this truth, and value the amazing opportunity we have to present it tomorrow. No matter the outcome, we have created together something beautiful and tomorrow will be a life experience that we will all share forever.

"I love working with this team. Knowing you all and creating together is a gift."

The warmth and intensity of emotion in that room sent tears down my cheeks, and I saw others with tears, as well. "Thank you all from the bottom of my heart for the intense and unremitting work you have all put in. Now, go back and get some dinner and some sleep, and tomorrow, let's DO THIS THING."

FIVE A.M., AND MY EYES FLEW OPEN IN THE DARK, BEFORE THE ALARM on my iPhone could chime. This was the day. The review panel was coming. I mentally jumped out of bed, but my body still couldn't do it, so I gently rolled onto my side and then hobbled into a hot shower. I was AWAKE. Out of the shower, I considered the two outfits I had brought. Which one. The choice depended on how I was feeling about my body. I lay down on the bed on my back in order to put on my black tights; I could not lean over to put them on my feet. Sigh. This damaged body, and its glacial rate of tiny improvements, so slight that I sometimes thought I was imagining getting better and really it was endlessly the same pain, the same dysfunction. No time to think about that.

Black Eileen Fisher skirt, black sleeveless blouse, wine-red blazer. White blouse and red blazer. No, black blouse. Short hair, freshly cut, my now natural gray and silver. And then, my special good-luck gift to myself for this event: black patent ankle boots with rectangular white jewels around the low boot heels. The most expensive shoes I had ever bought, by far, and I didn't even believe in the concept of good luck. But I did believe in props that made me feel strong.

Down through the quiet hotel, into the garage, and over the dark streets in my rental car to the Maxar parking lot and my assigned space. Even the parking was carefully orchestrated. Once it was light out, there would be people and signage to get the review panel members into the most convenient parking spots.

There across the front of the building was the huge banner with our Step 1 Concept Study Report cover art, a blazing impact between two planetesimals, one streaking away in blue and the other with its crumbling rock revealing the incandescent liquid metal core, and "PSYCHE" across the bottom. I swiped myself in with my badge. Inside stood the welcome desk with every review panel member's badge. Across from the desk was the two-meter-high model of the asteroid that Maxar had had built and painted, and around it, pedestals with meteorites from the Smithsonian National Collection that showed the story of Psyche: metal, and rock, and gleaming green olivine and pyroxene crystals. Tim McCoy, one of our science coinvestigators for the project, and the curator of the collection, stood guard. Tim and I said hello, just a word, what else to say?

Mainly, though, the building was quiet: the team was not due until 6:45. I walked up the stairs and past the small lobby, where a wide-screen monitor was already looping our computer-generated video of the spacecraft and its transit via our artist's interpretation of what Psyche might look like; artist Peter Rubin and I had spent Sundays for over a year to create this melding of science and art. The long hallway to the meeting room had vertical decals with "Psyche" in our branded orange and purple marching down the walls, along with displays and art about the mission and the asteroid, and Psyche-branded signs to the restroom, coffee, lunchroom, and the presentation room.

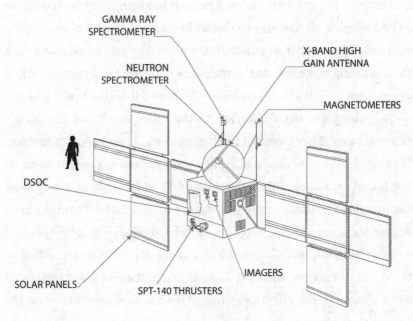

Diagram of the Psyche spacecraft.

I walked down this long quiet hallway, feeling the moment. It was about to be showtime, but for now, it was quiet, it was still *before*. I opened the door to the presentation room. I was still alone. I walked to the top left edge of the U of tables closest to the screen. Our team leadership would sit at these arms of the U, closest to the presenters, always visible to the board members, who would sit along the bottom of the U and in several more rows of broad clean tables. All the tables were set to perfection with water, glasses, and tented name tags with the Psyche logo at each place, exactly where we wanted each person; actual hours of conversation had gone into the seating by the team leadership and the review panel. Every chair, microphone, timer, window blind, and bottle of water was exactly where we planned it should be.

I pulled out my laptop and assembled a few things I would need; my phone, charging cables, and pens. The throat lozenges that Cami Skiba, my amazing colleague at ASU, had sent me in a complete Site Visit Care Kit she assembled not only with the lozenges, but also candy and pencils. I opened my laptop. I needed to arrange my desk, prepare to be looked at all day, run to the bathroom, get some coffee, but the email was there and it was all that mattered. The encrypted email with the questions that arose as the reviewers had read the material we had given them as PDFs two days before. I opened it with a beating heart.

All along I've been referring to the "questions" that the review board gave us. They were not really questions. Their proper name was "Potential Major Weaknesses." They were statements, or findings, by the board on issues which, if we couldn't address them, could tank the mission. On Day 1 of the site visit we received 31 of them: 10 to be responded to in writing two days before site visit, 9 to be responded to in writing the morning of the site visit, and 12 to be presented to the panel during the site visit. One example that we'd worked on all week was about the effects Psyche's magnetic field might have on the space around it:

C.15   The CSR anticipated the existence of a Psychean magnetosphere but did not address the risk associated with operating the spacecraft in a space environment where the associated energetic charged particles potentially result in Total Ionizing Dose (TID), Displacement Damage Dose (DDD), and peak charged particle flux requirements significantly higher than proposed. The CSR stated that "Psyche will likely be the most magnetic planetary object ever encountered at close range by a deep space spacecraft (summary in [Richter et al., 2012]; Psyche's field may be as large as one-half of the Earth's) . . . Psyche will also be the first asteroid encountered that could generate a magnetosphere." However, the CSR did not discuss the effect of the energetic particles that could be produced in the resulting solar wind-magnetosphere interaction, the magnetosphere's energization processes' effect, or the interaction of all space radiation environments with Psyche and its magnetic field on the expected local charged particle environment. While the CSR included a discussion of radiation effects, it did not address the impact of these additional potential factors on the science and mission implementation. Major hardware redesigns, including use of radiation hardened parts, throughout the instruments and flight system may be required and the overall mission might have to be redesigned in order to meet the science objectives. This finding represents a cost threat assessed to have a Likely likelihood of a Limited cost impact during development [Sec. E.2.2.1; Fig. F.4–2, Sections F.3, G.2, and G.3].

We had added spice to the Concept Study Report by discussing the possibility that Psyche would retain the magnetic field of its early active age. The area around Psyche dominated by its magnetic field, rather than the Sun's, is its magnetosphere. True, it was possible Psyche would have a strong magnetic field! But it may have a small field, or no field. Of course if it had a strong field, the reviewers were pointing out, the field could accelerate charged particles from the solar wind and those particles could damage the spacecraft. The spacecraft could be effectively flying through a particle accelerator. The unknown aspects of this unknown asteroid were suddenly a risk to our mission, not just a compelling reason to explore. We'd worked all week to make calculations and models to address this magnetosphere risk.

And now, in this email, were . . . thankfully, just three more Potential Major Weaknesses. I heard the greeters at the front door, way down the hallway and down the broad staircase into the lobby with the giant model of Psyche and the meteorites from the Smithsonian, begin to call out hellos to the many team members as they arrived. The suspended quiet of the morning, when all this was still a theoretical future, was over. Suddenly there was not enough time and I had to rush. Bob Mase, our deputy project manager, came by and we briefly discussed how the three new questions would be triaged: the lead responders would be Luke Dubord, Mike Ravine, and Brian Johnson. We lined up two people each to be first reviewers of the material, and one person for each question to be second reviewer. The presenters, later in the day, would be Mark Brown, Jim Bell, and Henry Stone. All this went into the spreadsheet that was accessible to everyone on the leadership and the Red Team. And now it was time to straighten my jacket and greet the team, all pouring into the room as Bob and I finished our update.

I know this all-team meeting happened but I have no memory of it. It was fifteen minutes stuck between the email and the review panel arriving. I walked down the long hallway, now with many people coming and going. I greeted friends with a special excitement and pleasure, amid the intensity of that moment, and then I began to see review panel members coming up the stairs.

Ours is a small world, and many of these panel members were friends, friendships from another, pre-competition time, suspended now. Out in the hallway, I shared a smile and silent acknowledgment of the importance of the moment as we passed with our payload (the science instruments) manager, Kalyani Sukhatme. The day before she and I had taken a couple of hours for a break at a salon. Now, it was game day, and I knew her technical skills were best-in-world, and she would give a great presentation later. The solidity of expertise in the team was calming. I could worry just for me, I didn't need to worry about them.

At the top of the stairs I could see down into the lobby with the meteorite model, and there was Tim McCoy, taking the review panelists through the meteorite collection and explaining the excitement, and the importance, and all the unknowns of our proposed exploration. Tim's depth of conviction was infectious and I could feel the energy and enthusiasm rising. I heard laughter, and questions, and I saw smiles. A good beginning. I turned and went back to the presentation room, and soon, it was filling up. Reviewers were filing in, searching for their name tags, greeting each other. And it was eight, time to begin, and the face of every reviewer was now serious.

For what is more serious in the world of commerce than a contract worth $800 million dollars? We academicians seldom experience the intensity of a competition where even the biggest companies straighten right up with interest and seriousness, where materials deemed "competition sensitive" are treated like top secret folios, where organizations are willing to invest millions simply to compete.

Henry Stone, graying hair combed and wearing a good suit over a freshly ironed white shirt, the heaviness of its fabric attesting to its quality, was now standing up front, lavaliere mic in place, welcoming the group and calling us all to order with his usual relaxed, regular-guy tone that as ever made us all lean in a bit and feel a part of what was about to happen. Jeff Marshall from Maxar explained the logistics of the building, entry and exit, security, coffee, restrooms. And Washito Sasamoto, the board chair, explained the process we'd go through that day. As Washito spoke I was putting on my own lavaliere mic. We had a couple of people making sure the mics were clipped on each speaker well ahead of their turn, and came off when they completed their turn.

I walked to the X in masking tape on the floor, the mark we put so that each speaker would stand out of the way of the projector but be visible to all reviewers. There they could see the monitors on the floor just in front of the panelists' tables, and I could see the slides that were showing on the screen behind me and never have to turn away from the reviewers. I could face them without a pause. I looked out over the reviewers, every eye on me, and I looked past them to the rest of the team filling the back of the room, and I felt wonderful. Alive.

"Hello, everyone. I'm Lindy Elkins-Tanton, PI of the Psyche mission."

The executive vice president from ASU made a welcome, and the director of JPL, and the executive vice president of Maxar, and then I was back on with a science overview. I felt easy in front of the room, talking about something I believed in so strongly: the inspiration and imperative of exploring new kinds of worlds. So easy to raise the energy of the room and point all our minds spaceward with the idea of this metallic body out there, waiting for us to come and discover what it is. I introduced the team and team structure, and pointed out key people around the room (I had memorized where each person would be sitting in the ranks of chairs behind the review panel). People were smiling and nodding. My shiny shoes were working for me.

And then, at 8:35, I started the presentations designed to answer the board questions with a statement about the time commitments from Jim Bell and myself (we would be available 100 percent for Psyche if needed and our institutions supported us in that). My turn was over. They had not grilled me on my leadership experience, as the mock panel had done; this panel mainly wanted to ensure I was available enough. I was. *Thank you,* and back around the table to sit down and let David Oh present on systems engineering. I took some deep breaths, and drank my coffee, and I smiled and nodded as David gave his brilliant presentation. What an art it is, I thought, to present information in a way that both answers questions and reassures the board emotionally that the person is knowledgeable, dependable, and expert.

The board asked questions, and the speaker passed them to whoever on the team was the expert for that answer or, in a few cases, asked that we hold that question for an answer later, in which case the Red Team took it and worked on it. Electronic messages flew back and forth among all of us on the team in the presentation room and those on the Red Team down the hall. How close to ready was that presentation for this afternoon from this morning's first questions? What needed to be reviewed?

Around ten-thirty, Ben Weiss stood up to present on the question of a Psychean magnetosphere. We'd worked on this answer during the week and thought it was pretty straightforward. The written response was only three pages long, and its summary was:

The spacecraft and measurements would be at risk only from energetic (greater than ~1 MeV) trapped particles in the vicinity of Psyche. This could only occur if Psyche has an energetic magnetosphere. Three factors control the existence of such a magnetosphere: surface magnetization, magnetic field geometry, and the size of the body. While Psyche could conceivably be as magnetic as some iron meteorites, plausible magnetic field geometries and Psyche's size together make it extremely unlikely that Psyche could produce a hazardous energetic particle environment. Therefore major hardware or mission redesigns are not needed.

As Ben presented, I could sense a restlessness on the panel. Some small questions and answers, and then, a definitive question from one panelist: "How rigorous are the calculations of the possibility of a charged particle being trapped in the magnetosphere? I am concerned we don't have all the possible energy levels addressed here." That question had to be postponed into the afternoon, as more calculations would need to be done offstage before results could be presented.

Suddenly we were actively worried about the threat of an energetic magnetosphere around Psyche, a hypothetical magnetosphere around an as-yet-unexplored and uncharacterized asteroid that, if we couldn't address these hypotheticals, would not be visited at all.

During our lunch break we all gathered in a break room and put some food on plates and then gathered around the one whiteboard hanging on a wall. On that tiny whiteboard, Chris, Ben Weiss, and Maria Zuber, a kind of dream team of space physics, gave us all a primer on the development of a magnetosphere around a small body, and how that magnetic field interacts with the solar wind. People in front crouched down or sat so that those of us standing in the back could see.

Chris began by drawing an asteroid with a magnetic field with charged particles striking it from the solar wind, and being spun around the magnetic field lines and trapped. "What's the energy range of solar wind particles?" one person asked. Chris and Maria wrote values on the board. "What's the possible magnetic field of the asteroid?" another person asked, and Ben wrote a series of values. Murmuring voices came from the back of the crowd as some people explained the problem to others or postulated problems and answers. Ben and Maria drew some force vectors on top of a solar wind particle. We began to argue about the possible shapes of the magnetic field lines around the asteroid. The big gaps in knowledge began to be delineated. And then Maria, Ben, and Chris went into a small conference room with their laptops and paper and pencils to drive toward some answers. They had about two hours to come up with the scientific answer and a compelling way to present it.

After lunch, we left the three of them and went back to the presentation room for a session on the Multispectral Imager, Psyche's set of dual cameras. Prior to Day 1 of the site visit, we had not expected any questions on the imager. Jim Bell, the lead of the imager build and science investigation, and the deputy principal investigator of the whole mission, is perhaps the world's leading expert on science investigations with cameras. The idea of "heritage" is central in space mission planning: if an instrument has been flown in space successfully before and we could build and use an identical one, then it's a very low-risk piece of technology for us. Our high-heritage camera had been built in largely the same way and flown before on several missions. And yet, we received a number of questions about the imager.

Jim gave the presentations we had prepared in response to the questions from Day 1. And then it was time for the deferred responses: answers to questions posed to presentations as they were being done today, but that we were not ready to answer off-the-cuff in the moment. My mind was focused on the sixth of these, about the Psyche magnetosphere, but first, Jim Bell had to present answers to a series of additional questions.

After working together for a few years, I knew Jim pretty well, and I knew both that he was a consummate professional and highly experienced presenter, and that he had a well-controlled temper. Jim had concluded the bulk of his presentation and now was receiving some surprisingly naive questions from one of the review panelists. At one point, David Oh leaned over to me and asked, "Does that question even make logical sense? Is that possible?" "No," I answered. "I don't think that math even works." Jim strolled back and forth in front of the panel, gazing down, apparently deep in thought. I was wondering how his mind was parsing this challenge: *I can't answer the question because it literally makes no sense. I can't point that out to the reviewer without making an enemy.* Finally, Jim said, "I'm not sure I completely understand your question. Could you ask it again in a different way?" And the meeting went on.

Later, colleagues of ours from ASU sitting in the back of the room told me they laid a $50 bet over whether or not Jim would lose it and start pelting the reviewer with the many issues of *Science* and *Nature* that had cover photos taken by Jim's cameras over the years. Of course, he did not, and we all survived.

And now it was time for Ben Weiss to give our answer to the challenging magnetosphere problem. *How bad could it be? Would the spacecraft be in danger?* Ben came out to the presentation area, wearing the suit that his partner had been kind enough to FedEx to him when to our surprise we found he was going to be presenting.

Ben's first slide came up, showing the request we were attempting to answer: provide a more detailed calculation on the possibility of trapped particles in the magnetosphere. Ben flipped to the next slide. The slide's title was "Energetic Particles Can Not Get Trapped in Any Psyche Magnetosphere," and just below, "Here is why." First, Ben showed the calculations that even with a uniform worst-case (strongest) Psychean magnetic field, protons above a certain energy could not be trapped because the spatial size of Psyche's field is too small. And then, could lower-energy protons that *were* trapped be energized sufficiently by the field to damage the spacecraft? Again, he showed that the small spatial size of Psyche's possible magnetosphere limited how much they could be accelerated.

Then, the next and final slide. "Trapped Electrons Cannot be Energized," the title announced. Everyone in the room was wide awake and utterly engaged in this conversation. If Ben, presenting brand-new calculations done over the lunch hour, could not convince the board that the spacecraft would be safe from accelerated particles and thus able to do its mission, they might on that basis alone kill the mission.

Ben began to walk the room through the three calculations on the slide. *The magnetosphere is too small for solar wind reconnection. The solar wind pressure is too weak for Fermi acceleration.* And the final, key point: Ben began to explain that betatron acceleration did not pose a threat. He pointed to the equation. "The ratio of the initial to final size of the magnetosphere cubed equals the ratio of the final to initial energy. So it's a reciprocal cube relationship. Compressing the magnetosphere causes the electron's energy to decrease." That last word was wrong. Ben should have said "increase." And all of us in the room were following so closely that many of us immediately called out, "Increase." Not just Maria and members of the Psyche team: people on the review board called out the right answer, as well! In that moment I knew we were all on the same side.

Ben clapped his hands to his head and groaned in agony and slumped into a squat for a moment before springing back up and declaring, "Increase!" And then he went on to show the maximum energy that could be given to an electron, and how it was too small to jeopardize our spacecraft.

At last, the room was quiet. I had answered a deferred question about Jim Bell's and my time commitments: we would be 100 percent available to the mission. Bob Mase had answered a question about eclipse calculations involving how long Psyche's shadow would lie across the spacecraft. Reviewers were leaning back in their chairs; it was four-thirty, and we were finished. I stood up for my concluding remarks. Michael New, the director of the Discovery Program at NASA headquarters, asked me, "By the way, Lindy, where did you get that logo you're using for the mission? It's really different than most logos." For a moment I was nonplussed. And then I realized, Michael was relaxed enough to ask this logo question, and therefore, we had been more or less a success today.

"Michael," I answered, "we hired a talented young branding designer in San Francisco, Michael Taylor, to design a logo that was bright and inviting and not another piece of hardware flying through the blackness of space. We wanted a logo that said, everyone can be involved. The logo is an abstract image of a planetesimal's core, under its enveloping layers of rock, and the colors are meant to indicate the energy of the processes and also the emotions of all of us while we do this adventure."

And with that, our site visit was over. We had a big dinner party, flew home, and settled in to wait for the verdict.

A SPACE MISSION ADDS A SCALE OF LONG-TERM COMMITMENT THAT most scientists, most people, are unfamiliar with. The proposal process might take six years, as ours did, and it might take sixteen years, and it might not end in selection. Then, the mission itself, from planning to building to flying and then exploring and learning, might take as little as a few years, as for some of the *Apollo* lunar missions once the cadence was high; or it could take decades, as the *New Horizons* mission to Pluto and beyond has done; or a half century, as for the *Voyager* missions, which have now flown out of our solar system and yet are still in touch. Any one of these missions could span the entirety of a scientist's career.

And yet they may be canceled at any moment before launch because of an unforeseen budget downturn, or the failure of a critical component; they might even explode on the launchpad. The work and emotional investment of years or even decades can end in a moment.

How can a person prepare for that kind of devastation? The conscious knowledge of its possibility is never enough. A friend of mine worked for years and got into the final round of selection for a big NASA competition, only to see another mission selected for flight and her own team suddenly disbanded. People she had worked with shoulder to shoulder for years were off to other projects, and work patterns that had been as ingrained as her morning routine were suddenly gone. The pain of loss after that degree of dedication and commitment was too painful to repeat, she told me; she would not be proposing again.

What is this ability to risk so much devastation? People say, in a facile modern take, "enjoy the journey" and "be more present." As a child, I learned some of those beautiful lessons from my father. We would tour the garden each morning to see whether the number of morning glories surpassed all prior records. We'd gleefully plan our favorite snacks to enjoy in front of the Giants' football game, as we cheered with decreasing urgency as they inevitably lost in that long fallow period they suffered.

In adulthood, though, I've developed a different metric: if you are sure of your values and your vision, then every unit of work, every day of effort, is a real piece of progress and of value in itself. I hope I enjoy each part, too, and am present for it, but I am more concerned that it has meaning. Just being a leaf isn't enough for me in the long run. There are too many imperatives in our world today.

This mantra carried me while I was fighting cancer, at the same time as I had just started my new directorship at ASU and was leading the Psyche team in writing the Step 1 proposal. I knew that if I died I would not have regret, because I had not wasted my time. I had been engaged every minute. I could look back and know that even then something had been built. The work itself, every brick in the road, each of the human relationships, had value. The drive to keep progressing and producing kept me focused and moving forward.

And I knew, after the site visit, that all we had done was worthwhile, even if we weren't selected. And so, we settled in to wait.

Chapter 12

AT THE END OF THE
MARATHON, A SPRINT

A t 4:08 P.M. on January 2, 2017, in western Massachusetts, the sun
was sinking behind the hill to the west that belongs to our neighbors Alice and Paul, who also built our house and shed. The clouds
were yellow and the sky was blue and that was a lot of color in this
otherwise white and gray winter landscape. This evening felt like the
end of a Sunday, when years ago, my dad and I would have some tea together and stay up to watch a movie, staving off the arrival of Monday.

Sometime in the coming week I and the other four principal investigators of NASA Discovery mission concepts would learn which
projects had been selected for flight. Might be one, might be two. Just
maybe, our Psyche mission to fly to and study a metal asteroid between Jupiter and Mars would be one of these.

I'd only been working on this project for five and a half years.
Some of my competitors had been through the process before with
the same ideas, and were coming up on a decade of trying to fly their
concept. Still, five and a half years. About 150 people had worked on
this concept with me thus far. We'd written about two thousand pages,

including the Step 1 and Step 2 proposals and all the written, edited, revised, formatted, and published answers to questions that came from NASA in between. We had art and models and videos and new scientific and engineering results because of all our efforts to understand how to get to the metal-rich asteroid Psyche and what we might find if we did, and how we could measure it and send the information back to Earth and understand it and interpret it for everyone in the world.

To say our hearts were in this project would be too facile, too trite. We lived and breathed it. We knew and loved each other and we knew each other's families and we had learned when to make a joke, and when to be quiet and let the other person work through a peak of frustration late at night after no rest for weeks. We had sweated through countless reviews and celebrated with numerous cakes and dinners the many intermediate successes that allowed us to get here, the ultimate intermediate success, the privilege to wait for the phone call.

That phone call would come to me, out there in western Massachusetts, on our landline. I would answer on the Bakelite phone that for decades had sat in my parents' house in Ithaca. And then I would get to tell the team, good news or bad. Sometime this week.

I JUST LOOKED UP AND IT WAS NOW DARK OUT, AND EIGHT AT NIGHT on January 4, 2017. The day had more than flown; it had evaporated.

Hundreds of people labored for years on the five final NASA mission concept studies: VERITAS, DAVINCI+, Lucy, NeoCam, and Psyche. Some people have worked their whole careers to win a mission. There was an ocean of heartbreak today. The three missions not selected were outstanding. I heard NASA officials say that this was one of the strongest Discovery panels ever. I knew principal investigators from other calls and other years who said the heartbreak was too much, they couldn't go through the process again. I hoped none of the other proposing PIs felt that way. It was a high-risk endeavor. It was not literally our lives at stake, as it was for other explorers, but it surely was our hearts.

I was sound asleep when my mobile phone rang. Though I had received lots of great advice via social media about how to relax the evening before (most involving alcohol), I'd happily gone off to bed, read for a while, and slept so soundly that the plowman clearing our long gravel driveway in the middle of the night didn't even waken me. James had spent hours during the week before helping me reason through how it would go if we learned we had not been selected for flight. I was afraid of feeling that I had let everyone down, let down the scientists on the team, and the engineers at Jet Propulsion Lab, and most especially, let down the team at Maxar, who had not been through a bidding process like this before and really had leaned in and spent resources and given their all. Was I sure we'd really done our best?

Truly, I felt we had. We had never skipped a hard decision or gone light on something that needed time. But I felt we had the luxury of writing exactly the proposal we really thought was right, because we were the underdogs. We were not going to win, I figured, so I could more successfully steer us away from marketing and toward writing about the mission we really thought was best. We were not going to win because I was a first-time principal investigator and the proposal was going through the process for the first time; generally these proposals need to go through the multiyear process of review and improvement a couple of times before they are tuned up to the point they are selectable.

Then, Maxar was a new prime partner for NASA. Maxar had not built spacecraft power systems and chassis for deep space before, only for Earth orbiting. But Maxar had built so many that they knew exactly what the spacecraft would cost, what it would weigh, how long it would take to build, so they could give us a firm, fixed price (as opposed to the normal "cost plus" contract of the aerospace world). Being brand-new to NASA prime contracts, NASA would give them extra scrutiny, and since they claimed to be able to build this spacecraft for much less than the expected value, the scrutiny would be additionally detailed, and the chance of our being selected proportionately lower.

Finally, there was our destination. We were proposing to visit an asteroid, but everyone in the community said it was time to select a Venus mission. So that night I'd gone to bed pretty comfortable in the knowledge that the next day I'd hear a no, and then we'd just start getting ready to go through another decade of waiting for the NASA call for proposals, and then fighting through the years of competition.

Because of how much was depending upon this phone call and the fact I was supposed to be awake for it, I was embarrassed to be sound asleep when the phone rang at 8:00 A.M. Thomas Zurbuchen, NASA associate administrator, said, "Oh, did I waken you?" and "I think you will be glad I did!" What a wonderful way to wake up. Then the cell coverage was lost, as it normally is up here in the Massachusetts hills. He called back. The connection failed. He called again, and I managed to slip in, "Call on the landline!" He called back on the landline.

My team and I always hoped and yearned to win, but I suppose I might have been overprepared for losing. After I hung up, I went downstairs and put some fresh wood in the woodstove, and put on the kettle for coffee, thinking, I need to call my husband, I need to call my son, I need to call my brother, and I need to call the team: Henry Stone and Jim Bell and Ben Weiss and David Oh and David Lawrence and Carol Polanskey and many other team members . . . and I need to call Michael Crow, and Jet Propulsion Laboratory director Mike Watkins, and Maxar lead Paul Estey . . . But instead, I turned the kettle off, closed the woodstove flue, put on my boots, and walked up into the woods.

The snow was soft from a night of rain and freezing rain, and the tree trunks were dripping dark with water. As usual I listened to the sound of my boots in the snow, and looked for animal tracks: one coyote, one fox, one squirrel, one vole, two deer. I walked up to a saddle between two hills and just stood looking at the trees in the silence. I always look for porcupines in trees and I never see them. I see porcupines walking on the ground often in the spring, out here in the forests, but in the winter I know they go up the trees to eat bark and I should be able to spot them, with the leaves all down, so I always look and I never see them. I didn't see any this morning. Then I walked back to the house and started calling people.

At the end of that day, after 30 phone calls, 5 interviews, 1 NASA press event, perhaps 500 emails, Facebook posts, tweets, and texts, I thought I was beginning to grasp the first 3 percent of the realization that we were really going to do a mission to Psyche. We were really going there to see it. My heart swelled up and I felt maudlin embarrassing sentiments . . . so grateful, so grateful, so grateful, and so eager to share this adventure with everyone.

That was the green light I received on the phone that morning: not a prize, not a membership, but something much deeper and richer—a chance to try for something really big. We had five years of completing the design and building the spacecraft, and if we could do it and do it right, we had a chance to launch it on a rocket. And if the launch worked, and the spacecraft worked, we could operate it and track it as it cruised out past Mars and to Psyche across three years of travel in the dark and the cold, and then we could learn what Psyche really is.

We learned, later on, that the review board and the people from NASA headquarters had noticed that our team functioned well together. We passed questions to the right expert for the answers, rather than the leadership trying to be the only ones who talked. We supported each other when we were standing and presenting, rather than sitting with judgmental crossed arms or snide side comments. And so we passed from striving to be selected to strive to build a spacecraft, with the validation that trying to make the hard choices correctly, bringing in good people and listening to them, and trying to be kind really did matter.

I PUT EACH FOOT IN TURN INTO A MACHINE WITH WHIRRING brushes and a vacuum to clean my shoes, and I stepped immediately onto a sticky mat on the floor, like wrong-side-up tape, to catch any dust or particles still on the bottoms of my shoes. That sticky mat lay in front of the door into the gowning room. In that gowning room, already cleaner than the shoe-cleaning room, I cleaned my iPhone so that I might be allowed to photograph the spacecraft. Then I put on shoe covers, a hair cover, a long sturdy plasticized gown that zipped and snapped up to my chin, an N95 mask with a surgical mask over it, a wrist strap with an electrical grounding plug attached to it by a springy coiling wire, and surgical gloves, pulled up over my cuffs. The goal: bring no FOD (foreign object debris) into the high bay. The connectors and circuits of the spacecraft must not be compromised by conductive dust, and seals must not get grit in them.

The irony with all this personal protective equipment was, of course, that we had survived the year of COVID and we had had our vaccines, and so now, we could afford to worry about mere dust.

I pushed my grounding plug into a special receptacle and put my fingers on a conductive bar to ground my body (a person and their clothes can carry kilovolts of static electricity, and Psyche's circuits could be damaged by such a jolt if that person ended up grounding their electricity through the spacecraft). Then, the most futuristic part: the air shower. I was enjoying this ritual to an embarrassing degree. One more sticky shoe mat and then into a small room like a bright white phone booth with two doors, its walls and ceiling pockmarked with nozzles. As directed, I held my hands up and my arms away from my body and I did my "air dance," moving my feet up and down and spinning in the blast of air from all sides. I air danced until the air shower turned off automatically after the prescribed amount of time, and then the second door unlocked and I was through, into the high bay.

This huge clean room is about seventy feet square, and taller even than its width, and it is very clean: certified to a cleanliness level of Class 10,000 means that there are fewer than 10,000 particles of 0.5 microns or larger in size per cubic meter of air volume. How good are the air filters? On a good air quality day outside, there might be 700,000 particles 2.5 microns in size in each cubic meter of air, and the filters bring them almost to zero inside the huge room; any particles larger than that are completely excluded.

I knew not to touch anything, and I knew not to go past the stanchions or the tape on the floor ("Psyche magnetic cleanliness zone"). And there she was, the great big spacecraft, being worked on, that morning, by a group of five people gowned, masked, gloved, and grounded just as I was. That day they were preparing to install the communications panel, in the run-up to turning on the compute element (the "brain") and having Psyche come alive, just a couple of weeks later.

All this because we are now in ATLO: Assembly, Test, and Launch Operations. We are in the final stage leading up to launch. The cadence of the day and the week have sped up. There are more surprises and palpably higher tension on the team. The spacecraft chassis had been delivered about a month before from Maxar to JPL, where I visited it today in the clean high bay.

The massive Psyche spacecraft, basically a huge cube, is attached by a meter-wide adapter ring on one of its faces to a dolly, a massive rollable stand that holds the spacecraft off the ground and can tip or rotate it, so that the engineers can reach any part they need.

I feel a kind of vibrating wonder, snapping me acutely into the present moment of this bright space with tools, and giant monitors showing plans and diagrams, and steel tables and plastic bins full of parts kitted in nonconductive plastic bags, and ladders and steps, and people, and then back into the past, a decade ago, when all of this was so unlikely and such a distant and vague dream. How did we make it to this point? It seems impossible that I am really in the room with the spacecraft.

And once again, as with so many milestones on this long path to space exploration, this was just a gateway to a more intense and real step: launch, and the beginning of operating this spacecraft as it flew to Psyche. Launch. From then on, we can't touch the spacecraft, can't physically repair anything, can only communicate using the Deep Space Network and all the codes and commands we have designed into the software. We'd made it to the final year before that moment. The year that had to hold the completion and success of everything.

WE HAVE EACH DAY PLANNED FOR THE NEXT FIFTEEN MONTHS OF ATLO leading up to launch. I'm back home again in Arizona, and today is Wednesday, one of a seemingly endless number of Wednesdays during the COVID lockdown. Every day for a year it's felt like Wednesday. I'm sitting at my green enameled metal desk, looking out the window at the prickly pear cactus, and the block of pressed bird seed in its shadow, staffed at the moment by a desert cottontail rabbit, a pair of Gambel's quail, and their almost unbearably cute family of tiny butter-and-chocolate-striped newly hatched chicks. But it's time to stop looking, and attend to my calendar: the day is beginning, and like all days until the launch of Psyche, it's going to be a long one.

In my Google Calendar the day is a seamless rainbow of meetings from now, 7:30 A.M., through 6:00 P.M. It doesn't have to be this way—even leading a mission like Psyche, it's possible to avoid being overscheduled—but working on the ASU Interplanetary Initiative and

Beagle Learning and Psyche means even the minimum set of good, important meetings fills the whole week. I still feel urgency about making progress on all the things that seem important, possibly scalable, possibly transformative. And so I click on the calendar invitation and open the Zoom link to the first meeting of the day, the Beagle founders' check-in.

Carolyn Bickers is on already from Boston, looking fresh and with her long hair smooth. She has on her game face and so I know she has some big sales efforts for the day. James comes on about the same time I do, from the other end of the house, where he likes to sit at our dining table, looking out into the backyard. We've been like this most of COVID, me shut into the study, him in the open living/dining/kitchen area. And now Turner is on, from New Hampshire, his dirty blond hair and beard framing his steady hazel eyes, his face lit by a big smile. "How is everyone?" he asks. Carolyn says she's fine, as she always does, and tells a funny story about her brother. I update everyone on the baby quail. And then on to the business of the day.

Our half hour is up and it's time for me to hop to the Psyche Systems Engineering Team meeting. We look at near-term engineering activities and a series of Engineering Change Requests—changes to the plan that need to be reviewed and discussed by all the subsystems that might be affected. Near the end of the PSET meeting I get a text from Henry Stone, our project manager. *Can you call me?* Uh-oh. These are really never good news. I text back, *In five minutes?* And Henry replies, *Copy.*

On the phone Henry tells me about a new slip in the schedule for one of our subsystems. The group responsible had ordered all the parts they needed to build the instrument ahead of time so the parts would be there and ready. But the parts were insufficiently inspected upon receipt, and they sat there for a couple of months in storage, and then when they were needed they were found to be faulty. Things we don't know yet: How long it will take to get the parts repaired, or replaced, if they must be. Whether the team building the instrument can re-organize their schedule to accommodate the delayed parts in such a way that the final completion and delivery of the subsystem isn't too much later than it is supposed to be. Henry explains all the facts in his thorough way, and then we talk about the human actions, as we always do. How can we create circumstances that will support the team to avoid repeats? Is there anyone who needs coaching? Our alignment on team culture feels near-miraculous and it has smoothed the road. Action taken: a meeting is scheduled with Kalyani Sukhatme and the subsystem lead and their team, to discuss process and schedule; they will give us daily updates, and will work on their schedule in partner-ship with the JPL project team.

With every day planned up to launch, schedule slips become se-rious indeed. The science instruments need to be installed on the spacecraft before the spacecraft goes through "environments": vibra-tion, acoustics, thermal-vacuum chamber, all the tests meant to show the spacecraft will work as planned after launch and in the space environment. If one event in the schedule, like the acceptance and integration of the science instrument, slips later, then something else has to be rearranged to happen sooner. And there are only so many things that can move sooner, so the more slips there are, the more brittle the schedule becomes.

I'm grateful the instrument team let us know immediately that something had happened. Immediate, complete communication is a requirement for being able to find solutions and work ahead when problems occur. There will always be problems; people will make mis-takes, and things will break. The goal is having a team that works well enough together to be able to absorb these problems, and to catch them in the first place.

Ten years ago, when we started pitching the Psyche mission concept to review teams inside JPL, we began to set our culture: We would give as clear and accurate a picture as we could of where we really were in our process. We would not spin, or obfuscate. We'd show what we'd accomplished, and we'd discuss our challenges and ask for input. And this has paid off so many times—in gaining trust, in getting help instead of reprimands, in being given the benefit of the doubt.

Keeping that culture going is a near-daily effort. We each have such an instinct to hide the mistakes, hide the embarrassing things. I struggle with it myself. Instead we take a deep breath and tell the whole story. Look for collegiality and advice, rather than anticipating punishment. Our deputy payload manager, Noah, called this team practice "radical transparency."

A cousin to this practice is our Twitter hashtag #PI_Daily. I'd heard so many curious voices wanting to learn more about being a PI, and mainly what's available are the stories of the resulting science, and of the launch, and new discoveries. What about the day-to-day? How can anyone else know if they want this job, when it's so specialized and hidden? So a number of PIs on Twitter got together, with NASA HQ's blessing, and we tweet about what our jobs really entail. We want the job to feel closer and more accessible to anyone interested, but particularly to those who don't already have an inside track and could use an introduction.

Every tweet, I find, is a balance between NASA HQ shiny perfect, and the absolute truth in all its grittiness, and needing to avoid insulting anyone by bringing up the real challenges of the day. Want to be called out in public for having broken a piece of flight hardware? That might be the biggest news of the day, and it would be great for potential PIs to hear about it, but it's not fair to the person who did it, especially given the vagaries of a world that might overreact and even affect stock prices, in some dramatic examples.

Through writing those tweets, and our Psyche blog pieces, I've realized that just as the mission goes through the stages of proposal, design, build, launch, cruise, and finally arrival at the target and generation of data and new scientific discoveries, so does the work of the PI go through phases entirely different one from the next. In the proposal stage, we're figuring out how to tell the compelling story of the mission and all it will discover, and we're building the team, making decisions on who to trust and work with for ten or twenty years into the future. Are most scientists or engineers trained how to do these things? No, we are not. Then, after selection, it's management, risks, leadership, presentation, organization, communication. The science doesn't come till later. You can see the challenge: The PI needs to spend about a decade becoming an expert in communications, negotiations, team building, leadership, management—mostly new things to us. And then, after that decade, we have to go back to being brilliant at science. And we can't have lost track of the field during that time: we have to know all the people, all the discoveries, all the new hypotheses, and how what we'll learn will advance the field, either proving or disproving the ideas of the time. I try to share those little revelations in my tweets.

And I fly to visit the team at Jet Propulsion Lab, too, now that COVID restrictions are easing and my presence can be allowed. The excitement is palpable. I visit the team in the test bed, where the young manager and three even younger engineers tour us through the racks of instruments, boards, and harnesses that are duplicates of all that will fly on the spacecraft. This team tests that the software communicates correctly, and the instruments respond correctly, and the data gets relayed correctly. They tell me about their COVID challenges: the people needed to operate the lab's antennas to test Psyche's communications were not in, because of severe restrictions on the number of people allowed in the lab. These test-bed engineers found hardware and managed to cobble it together to make some preliminary tests before the regular transmissions were ready, and they kept the test-bed schedule going that way. They laugh about this story, saying, "It looked like the right kind of connector and the right power level, and we tested it, and it worked for us!" and suddenly their thrill of troubleshooting and making it all happen against the odds is infectious, and my worry falls away, and I laugh, too.

We need this, going forward to launch; we need to laugh and feel our personal power and the power of our team to overcome the challenges that are coming our way. It's not just for our space mission, our audacious attempt to send a robot on a path through space 2.4 billion kilometers long to a small metallic world that we humans just have to know about—we need to laugh and feel that power in all we do, and in all the challenges we face. So laugh with us, and feel that power, because we need it all to get to our launch on that big wild SpaceX Falcon Heavy rocket, just a few months away now. And then, with that launch and the start of Psyche's cruise through space, we'll have won, again, something actually worth winning: the chance to work harder, for longer, on something that will amaze us and drive human knowledge farther.

ACKNOWLEDGMENTS

Because I've written just a portrait, a sketch, these storylines do not encompass all the friends who bring that bloom of warmth in my chest and create a world that has safety and comfort: Dave and Erin Beaudet; Carolyn and Michael Bickers; Sara, Tom, Jared, Natalie, and Lu Collina; Ellie Doris and Bob, Micky, and Kathleen Strachota; Anke Friedrich; Mary Fuller; Tanya Furman; Andrea Hammer and Amy Villarejo; Jackie Mow; Sarah Ploss and Lap Wong; Susan and Steve Potter; Andie Shauger and Tim Ikeda; Evgenya Shkolnik and Aaron Dragushan; Caroline Smith and Jessica Robinson; Henry Stone and Juliana Murphy; Ben Weiss and Tanja Bosak; Gregg and Deb Vane; Alice and Paul Vigliani; Sue Webb; B. J. and Ellis Wiggins. Your kindnesses and companionship over the many years is more than I deserve.

Having a colleague and mentor who is also a friend is a kind of good fortune that opens wide horizons: Maria Anguiano, Steve Battel, Sam Bowring, Michael Crow, Valeri Fedorenko, Wendy Freedman, Nancy Gonzales, Tim Grove, Brad Hager, Paul Hess, Marc Parmentier, Vera Rubin, Everett Shock, Sean Solomon, Ellen Stofan, Nafi Toksöz, Maria Zuber, Thomas Zurbuchen.

Oh, team Psyche! You feel like a family to me, and like a big family, are too many to list, though we are together in this great adventure. These leaders have all had a critical role in the mission's development, and are valued colleagues and friends: Henry Stone, Carl Adams, Deborah Bass, Jim Bell, Bobby Braun, Diane Brown, Mark Brown, Richard Cook, Tracy Drain, Howard Eisen, Charles Elachi, Paul Estey, Sabrina Feldman, Lori Glaze, General Larry James, Gentry Lee, Bob Mase, Michael New, Sarah Noble, David Oh, Keyur Patel, Carol Polanskey, Kim Reh, Steve Scott, Brent Sherwood, Kalyani Sukhatme, Mike Watkins, Belinda Wright, Jakob van Zyl.

For years I'd wished to write down my story. Jane von Mehren at Aevitas (thank you for the introduction, Steve Beschloss) worked so supportively, lightly, and clearly with me over the initial writing and proposal that my confidence grew with the editing. This was a new experience of editing for me. And then Nick Amphlett at HarperCollins saw the vision and, miraculously, guided my efforts so that I learned to be a better writer and enjoyed the process and the outcome, as well. Chris Vasquez, a former Psyche Inspired art intern, created the spacecraft image for this book.

To Drs. Javier Magrina and John Camoriano, who saved my life and then worked on making it really good again.

And as I wrote this book, the times I have hurt people and left unsolved pain behind were always present in my mind. So especially to Theo, Deb, Ed, and Sharon, I am sorry.

Best for last. James Tanton, Turner Bohlen, and Liz Casey, you shine in my heart every day and are the structure and meaning of my world. How blessed I am to have family: Jim Elkins and Margaret MacNamidhe; Jeffrey, Wendy, Katherine, and Alex Cohen; Curtis Bohlen and Caroline, Sarah, and Felix Norden; Julie, David, Eliza, and Sophie Perry; K, Eleanor, Jen, Scott, Eric, and Andrew Isdaner; Art, Eleanor, and Sarah Kahn; John and Cathy Tarbox; Nina S. Bohlen and Robert Williams; Nina Bohlen; Bob and Shirlene Elkins; Marshall Elkins and Melissa Warlow; Fletcher and Abby Tanton; Sally and Leonard and Tom Elkins; Buff and Janet Bohlen.